Advances in
MATHEMATICAL ECONOMICS

Aims and Scope. The project is to publish *Advances in Mathematical Economics* once a year under the auspices of the Research Center forMathematical Economics. It is designed to bring together those mathematicians who are seriously interested in obtaining new challenging stimuli from economic theories and those economists who are seeking effective mathematical tools for their research.

The scope of *Advances in Mathematical Economics* includes, but is not limited to, the following fields:

- Economic theories in various fields based on rigorous mathematical reasoning.
- Mathematical methods (e.g., analysis, algebra, geometry, probability) motivated by economic theories.
- Mathematical results of potential relevance to economic theory.
- Historical study of mathematical economics.

Authors are asked to develop their original results as fully as possible and also to give a clear-cut expository overview of the problem under discussion. Consequently, we will also invite articles which might be considered too long for publication in journals.

More information about this series at http://www.springer.com/series/4129

Shigeo Kusuoka • Toru Maruyama
Editors

Advances in Mathematical Economics

Volume 20

Springer

Editors
Shigeo Kusuoka
Professor Emeritus
The University of Tokyo
Tokyo, Japan

Toru Maruyama
Professor Emeritus
Keio University
Tokyo, Japan

ISSN 1866-2226 ISSN 1866-2234 (electronic)
Advances in Mathematical Economics
ISBN 978-981-10-0475-9 ISBN 978-981-10-0476-6 (eBook)
DOI 10.1007/978-981-10-0476-6

Printed on acid-free paper

This Springer imprint is published by Springer Nature
The registered company is Springer Science+Business Media Singapore Pte Ltd.

Preface

This volume of *Advances in Mathematical Economics* is basically a collection of selected papers presented at the 6th Conference on Mathematical Analysis in Economic Theory, which was held in Tokyo January 26–29, 2015. The conference was organized by the Japanese Society for Mathematical Economics. On behalf of the organization committee of the conference, we would like to extend our cordial gratitude to the Keio Economic Society, the Top Global University Project, and the Oak Society, Inc. for their generous financial support. And of course, it is a great pleasure for us to express our warmest thanks to all the participants of the conference for their contributions to our project.

Professor C. Castaing's paper and T. Maruyama's paper were not read at the conference but are included in this issue.

This volume contains the whole programme as well as some photographs taken on this occasion.

December 14, 2015

S. Kusuoka, T. Maruyama
Managing Editors of *Advances in Mathematical Economics*
General Managers of the 6th Conference

We formerly classified our international research meetings into two categories, "conference" and "workshop". However, we relinquish this distinction from now on for the sake of simplicity and just use the term "conference".

Contents

Part I
Research Articles

Adv. Math. Econ. 20, 3–22 (2016)

Advances in
MATHEMATICAL
ECONOMICS

Local Risk-Minimization for Barndorff-Nielsen and Shephard Models with Volatility Risk Premium

Takuji Arai

Abstract We derive representations of locally risk-minimizing strategies of call and put options for Barndorff-Nielsen and Shephard models: jump type stochastic volatility models whose squared volatility process is given by a non-Gaussian Ornstein-Uhlenbeck process. The general form of Barndorff-Nielsen and Shephard models includes two parameters: volatility risk premium β and leverage effect ρ. Arai and Suzuki (Local risk minimization for Barndorff-Nielsen and Shephard models. submitted. Available at http://arxiv.org/pdf/1503.08589v2) dealt with the same problem under constraint $\beta = -\frac{1}{2}$. In this paper, we relax the restriction on β; and restrict ρ to 0 instead. We introduce a Malliavin calculus under the minimal martingale measure to solve the problem.

Keywords Locally risk-minimizing strategy • Barndorff-Nielsen and Shephard models • Stochastic volatility models • Malliavin calculus • Lévy processes

---

JEL Classification: G11, G12

Mathematics Subject Classification (2010): 91G20, 60H07

T. Arai (✉)
Department of Economics, Keio University, Tokyo, Japan
e-mail: arai@econ.keio.ac.jp

S. Kusuoka, T. Maruyama (eds.), *Advances in Mathematical Economics Volume 20*, Advances in Mathematical Economics,
DOI 10.1007/978-981-10-0476-6_1

Article Type: Research Article
Received: June 4, 2015
Revised: December 12, 2015

1 Introduction

Locally risk-minimizing (LRM, for short) strategies for Barndorff-Nielsen and Shephard models (BNS model, for short) are discussed. Here local risk-minimization is a very well-known quadratic hedging method of contingent claims for incomplete financial markets. On the other hand, BNS models are stochastic volatility models suggested by Barndorff-Nielsen and Shephard [2, 3]. It is known that some stylized facts of financial time series are captured by BNS models. The square volatility process σ^2 of a BNS model is given as an Ornstein-Uhlenbeck process driven by a subordinator without drift, that is, a nondecreasing pure jump Lévy process. Thus, σ^2 is a jump process given as a solution to the following stochastic differential equation (SDE, for short):

$$d\sigma_t^2 = -\lambda\sigma_t^2 dt + dH_{\lambda t}, \quad \sigma_0^2 > 0,$$

where $\lambda > 0$, H is a subordinator without drift. Now, we denote by S the underlying asset price process. The general form of S is given by

$$S_t = S_0 \exp\left\{ \int_0^t \left(\mu + \beta\sigma_s^2\right) ds + \int_0^t \sigma_s dW_s + \rho H_{\lambda t} \right\},$$

where $S_0 > 0$, μ, $\beta \in \mathbb{R}$, $\rho \leq 0$, W is a 1-dimensional Brownian motion. The last term $\rho H_{\lambda t}$ accounts for the leverage effect; and $\beta\sigma_s^2$ is called the volatility risk premium, which is considered as the compensation required by investors holding volatile assets. From the view of (2) below, the volatility risk premium vanishes when $\beta = -\frac{1}{2}$. So that, β would take a value greater than or equal to $-\frac{1}{2}$. For more details on BNS models, see Cont and Tankov [4] and Schoutens [11].

Our purpose is to obtain representations of LRM strategies of call and put options for BNS models under constraint $\rho = 0$ and no constraint on β. On the other hand, Arai and Suzuki [1] studied the same problem under constraint $\beta = -\frac{1}{2}$ and no constraint on ρ. That is, they dealt with the case where volatility risk premium is not taken into account. To the contrary, we will treat BNS models with volatility risk premium. In other words, we relax the restriction on β. Instead, we restrict ρ to 0, which induces the continuity of S. Then, S is written as

$$S_t = S_0 \exp\left\{ \int_0^t \left(\mu + \beta\sigma_s^2\right) ds + \int_0^t \sigma_s dW_s \right\}. \tag{1}$$

Actually, the continuity of S makes the problem easy to deal with. To calculate LRM strategies, we need to consider the minimal martingale measure (MMM, for short). When S is continuous, the subordinator H remains a Lévy process even under the MMM. On the other hand, the generalization of β makes the problem complicated. When $\beta = -\frac{1}{2}$, the density process Z of the MMM is given as a solution to an SDE with the Lipschitz continuity. Thus, as shown in [1], Z has the Malliavin differentiability under $\mathbb{P}$, which played a vital role in [1]. However, this property is not generalized to the case of $\beta \neq -\frac{1}{2}$. Hence, we need to take a different approach from [1]. In order to overcome this difficulty, making the best of the fact that the Lévy property of H is preserved, we innovate a Malliavin calculus under the MMM. As a result, we can calculate LRM strategies without attention to the property of Z.

To our best knowledge, except for [1], there is only one preceding research on LRM strategies for BNS models: Wang, Qian and Wang [15]. Besides they treated the problem under the same parameter restrictions as ours, although they did not use Malliavin calculus. However, their discussion seems to be inaccurate mathematically. For example, they did not mention the SC condition defined in Sect. 2.1 of this paper, any condition on the Lévy measure of H, and any property of the MMM. These are indispensable to discuss LRM strategies. In addition, their expression of LRM strategies is given by a partial derivative of a function on t, S_t and σ_t. It is very difficult to calculate it explicitly. On the other hand, we give a concrete representation for LRM strategies by using Malliavin derivative of put options.

Outline of this paper is as follows. A precise model description and standing assumptions are given in Sect. 2. In Sects. 2.1, 2.2, and 2.3, we define an LRM strategy, the MMM and a Malliavin derivative, respectively. Our main results are provided in Sect. 3; and conclusions will be given in Sect. 4.

2 Preliminaries

We consider a financial market model in which only one risky asset and one riskless asset are tradable. For simplicity, we assume that the interest rate is given by 0. Let $T > 0$ be the finite time horizon. The fluctuation of the risky asset is described as a process S given by (1). We consider a complete probability space $(\Omega, \mathscr{F}, \mathbb{P})$ with a filtration $\mathbb{F} = \{\mathscr{F}_t\}_{t\in[0,T]}$ as the underlying space. Suppose that $\mathbb{F}$ is generated by W_t and $H_{\lambda t}$; and satisfies the usual condition, that is, $\mathbb{F}$ is right continuous, and $\mathscr{F}_0$ contains all null sets of $\mathbb{P}$. The asset price process S given in (1) is a solution to the following SDE:

$$dS_t = S_{t-}\left\{\mu dt + \hat{\beta}\sigma_t^2 dt + \sigma_t dW_t\right\}, \tag{2}$$

where $\hat{\beta} := \beta + \frac{1}{2}$. Denoting $A_t := \int_0^t S_{s-}\left[\mu + \hat{\beta}\sigma_s^2\right] ds$ and $M_t := S_t - S_0 - A_t$, we have $S_t = S_0 + M_t + A_t$, which is the canonical decomposition of S. Further, we denote $L_t := \log(S_t/S_0)$ for $t \in [0, T]$, that is,

$$L_t = \mu t + \beta \int_0^t \sigma_s^2 ds + \int_0^t \sigma_s dW_s.$$

Defining $J_t := H_{\lambda t}$, we denote by N the Poisson random measure of J, that is, we have $J_t = \int_0^\infty xN([0,t], dx)$. Denoting by ν the Lévy measure of J, we have that $\widetilde{N}(dt, dx) := N(dt, dx) - \nu(dx)dt$ is the compensated Poisson random measure. Remark that N and ν are defined on $[0, T] \times (0, \infty)$ and $(0, \infty)$, respectively; and $\nu(dx) = \lambda \nu^H(dx)$, where ν^H is the Lévy measure of H. Moreover, Proposition 3.10 of [4] implies

$$\int_0^\infty (x \wedge 1)\nu(dx) < \infty. \tag{3}$$

We need to impose the following standing assumptions on ν as in [1]. As stated in Remark 2.2 below, the standing assumptions do not exclude representative examples of BNS models, although parameters are restricted.

Assumption 2.1.

(A1) The Lévy measure ν is absolutely continuous with respect to the Lebesgue measure on $(0, \infty)$.

(A2) There exists a $\kappa > 0$ such that

- $\kappa > [2\hat{\beta}^+ + 1]\mathscr{B}(T)$,
- $\kappa \geq \hat{\beta}^2 \mathscr{B}(T)$, *and*
- $\int_1^\infty e^{2\kappa x}\nu(dx) < \infty$,

where $\mathscr{B}(t) := \int_0^t e^{-\lambda s} ds = \frac{1-e^{-\lambda t}}{\lambda}$ *for* $t \in [0, T]$.

Remark 2.2.

1. When $\beta = -\frac{1}{2}$, that is, $\hat{\beta} = 0$, (A2) is equivalent to the existence of $\varepsilon > 0$ such that $\int_0^\infty e^{(2+\varepsilon)\mathscr{B}(T)x}\nu(dx) < \infty$. In [1] dealing with the case of $\beta = -\frac{1}{2}$, $\int_0^\infty e^{2\mathscr{B}(T)x}\nu(dx) < \infty$ is assumed in their Assumption 2.2, which is almost the same as the above (A2) for $\beta = -\frac{1}{2}$.
2. We do not need to assume conditions corresponding to the second condition of Assumption 2.2 in [1], which ensures the positivity of the density of the MMM defined below, since the continuity of S implies that the MMM becomes a probability measure automatically.
3. Condition (A2), together with Proposition 3.14 of [4], ensures $\mathbb{E}[e^{2\kappa J_T}] < \infty$, which obviously includes $\mathbb{E}[j_T^2] < \infty$.
4. Condition (A1) guarantees Assumption Z1 in Nicolato and Venardos [6], which we need in the proof of Lemma 2.9 below.
5. Assumption 2.1 does not exclude two representative examples of σ^2, "IG-OU" and "Gamma-OU". "IG-OU" is the case where ν^H is given as

$$\nu^H(dx) = \frac{a}{2\sqrt{2\pi}} x^{-\frac{3}{2}}(1 + b^2 x)e^{-\frac{1}{2}b^2 x}\mathbf{1}_{(0,\infty)}(x)dx,$$

where $a > 0$ and $b > 0$. The invariant distribution of σ^2 follows an inverse-Gaussian distribution with $a > 0$ and $b > 0$. Then σ^2 is called an IG-OU process. If

$$\frac{b^2}{2} > 2\left\{[2\hat{\beta}^+ + 1] \vee \hat{\beta}^2\right\}\mathscr{B}(T),$$

then Assumption 2.1 is satisfied. Next, "Gamma-OU" is the case where the invariant distribution of σ^2 is given by a Gamma distribution with $a > 0$ and $b > 0$. In this case, ν^H is described as

$$\nu^H(dx) = abe^{-bx}\mathbf{1}_{(0,\infty)}(x)dx.$$

As well as the IG-OU case, Assumption 2.1 is satisfied if

$$b > 2\left\{\left[2\hat{\beta}^+ + 1\right] \vee \hat{\beta}^2\right\}\mathscr{B}(T).$$

For more details on this topic, see also [6] and [11].

2.1 Local Risk-Minimization

In this subsection, we define an LRM strategy. To this end, we define the SC condition firstly; and show that S satisfies it under Assumption 2.1. S is said to satisfy the SC condition, if the following three conditions hold:

(a) $\left\| [M]_T^{1/2} + \int_0^T |dA_s| \right\|_{L^2(\mathbb{P})} < \infty$.
(b) A is absolutely continuous with respect to $\langle M\rangle$, i.e. $A = \int \Lambda d\langle M\rangle$ for some process Λ.
(c) The mean-variance trade-off process $K_t := \int_0^t \Lambda_s^2 d\langle M\rangle_s$ is finite, that is, K_T is finite $\mathbb{P}$-a.s.

Proposition 2.3. *S satisfies the SC condition under Assumption 2.1.*

Proof. First of all, we show item (a). We prepare the following lemma:

Lemma 2.4. *$\int_0^T \sigma_t^2 dt \in L^n(\mathbb{P})$ for any $n \geq 1$.*

Proof of Lemma 2.4. Since we have

$$\begin{aligned}\int_0^t \sigma_s^2 ds &= \sigma_0^2 \int_0^t e^{-\lambda s} ds + \int_0^t \int_0^s e^{-\lambda(s-u)} dJ_u ds \\ &= \sigma_0^2 \mathscr{B}(t) + \int_0^t \int_u^t e^{-\lambda(s-u)} ds dJ_u = \sigma_0^2 \mathscr{B}(t) + \int_0^t \mathscr{B}(t-u) dJ_u \\ &\leq \sigma_0^2 \mathscr{B}(t) + \mathscr{B}(t) J_t \leq \sigma_0^2 \mathscr{B}(T) + \mathscr{B}(T) J_T \qquad (4)\end{aligned}$$

for any $t \in [0, T]$, it suffices to show $J_T \in L^n(\mathbb{P})$ for any $n \geq 1$. By Remark 2.2, we have $\mathbb{E}[\exp\{2\kappa J_T\}] < \infty$, from which $J_T \in L^n(\mathbb{P})$ follows for any $n \geq 1$. □

Note that we have

$$\begin{aligned}
&\left\| [M]_T^{1/2} + \int_0^T |dA_t| \right\|_{L^2(\mathbb{P})}^2 \\
&\leq 2\mathbb{E}\left[[M]_T + \left(\int_0^T |dA_t| \right)^2 \right] \\
&\leq 2\mathbb{E}\left[\int_0^T S_{t-}^2 \sigma_t^2 dt + \left(\int_0^T S_{t-} |\mu + \hat{\beta}\sigma_t^2| dt \right)^2 \right] \\
&\leq 2\mathbb{E}\left[\sup_{0\leq s\leq T} S_s^2 \left\{ \int_0^T \sigma_t^2 dt + \left(|\mu| T + |\hat{\beta}| \int_0^T \sigma_t^2 dt \right)^2 \right\} \right].
\end{aligned}$$

If $\sup_{0\leq s\leq T} S_s \in L^{2a}(\mathbb{P})$ holds for a sufficiently small $a > 1$, item (a) holds by the Hölder inequality and Lemma 2.4.

Now, we take an $a > 1$ such that

$$\left\{ 2\left(a\beta + \frac{a^2}{2} \right)^+ + a^2 \right\} \mathscr{B}(T) < \kappa. \tag{5}$$

Note that we can find such an $a > 1$ from the view of (A2) in Assumption 2.1. We shall see $\sup_{0\leq s\leq T} S_s \in L^{2a}(\mathbb{P})$. Equation (4) implies that

$$\begin{aligned}
e^{aL_t} &= \exp\left\{ a\mu t + a\beta \int_0^t \sigma_s^2 ds + a \int_0^t \sigma_s dW_s \right\} \\
&= \exp\left\{ a\mu t - \frac{a^2}{2} \int_0^t \sigma_s^2 ds + a \int_0^t \sigma_s dW_s + \left(a\beta + \frac{a^2}{2} \right) \int_0^t \sigma_s^2 ds \right\} \\
&\leq C \exp\left\{ -\frac{a^2}{2} \int_0^t \sigma_s^2 ds + a \int_0^t \sigma_s dW_s \right. \\
&\qquad \left. + \int_0^t \int_0^\infty bx\widetilde{N}(ds, dx) + \int_0^t \int_0^\infty [bx + 1 - e^{bx}]\nu(dx)ds \right\} \\
&=: CY_t^{a,b},
\end{aligned}$$

where $b := \left(a\beta + \frac{a^2}{2} \right)^+ \mathscr{B}(T)$, and $C := \exp\{a|\mu|T + b\sigma_0^2 + \int_0^T \int_0^\infty (e^{bx} - 1) \nu(dx)dt\}$. Taking into account (5) and (A2) in Assumption 2.1, Lemma 2.5 below yields that $Y^{a,b}$ is a square integrable martingale. Thus, Doob's inequality yields

$$\mathbb{E}\left[\sup_{0\leq s\leq T} S_s^{2a}\right] = \mathbb{E}\left[S_0^{2a}\sup_{0\leq s\leq T} e^{2aL_s}\right] \leq S_0^{2a}C^2\mathbb{E}\left[\sup_{0\leq s\leq T}(Y_s^{a,b})^2\right]$$
$$\leq 4S_0^{2a}C^2\mathbb{E}[(Y_T^{a,b})^2] < \infty.$$

Next, taking $\Lambda_t := \frac{1}{S_{t-}}\frac{\mu+\hat{\beta}\sigma_t^2}{\sigma_t^2}$, we have item (b). Moreover, we see item (c). Noting that σ_t^2 is represented as

$$\sigma_t^2 = e^{-\lambda t}\sigma_0^2 + \int_0^t e^{-\lambda(t-s)}dJ_s,$$

we have that

$$\sigma_t^2 \geq e^{-\lambda t}\sigma_0^2 \geq e^{-\lambda T}\sigma_0^2 \tag{6}$$

for any $t \in [0, T]$. Thus, together with Lemma 2.4, item (c) follows. □

Lemma 2.5. *For $a \in \mathbb{R}$ and $b \geq 0$, we denote*

$$Y_t^{a,b} := \exp\left\{-\frac{a^2}{2}\int_0^t \sigma_s^2 ds + a\int_0^t \sigma_s dW_s + \int_0^t\int_0^\infty bx\widetilde{N}(ds,dx)\right.$$
$$\left.+\int_0^t\int_0^\infty [bx + 1 - e^{bx}]\nu(dx)ds\right\}.$$

1. *If a and b satisfy*

$$\int_1^\infty \exp\left\{\left(2b + \frac{a^2}{2}\mathscr{B}(T)\right)x\right\}\nu(dx) < \infty, \tag{7}$$

   then the process $Y^{a,b}$ is a martingale.
2. *When we strengthen (7) to*

$$\int_1^\infty \exp\{(4b + 2a^2\mathscr{B}(T))x\}\nu(dx) < \infty, \tag{8}$$

   $Y^{a,b}$ is a square integrable martingale.

Proof.

1. Theorem 1.4 of Ishikawa [5] introduced a sufficient condition for $Y^{a,b}$ to be a martingale, which condition consists of the following three items:

   (a) $\int_0^\infty [b^2x^2 + (1-e^{bx})^2]\nu(dx) < \infty$,
   (b) $\int_0^\infty [e^{bx}\cdot bx + 1 - e^{bx}]\nu(dx) < \infty$, and
   (c) $\mathbb{E}\left[\exp\left\{\frac{a^2}{2}\int_0^T \sigma_t^2 dt\right\}\right] < \infty$.

   By (3) and (7), items (a) and (b) are satisfied. Next, (7) and Proposition 3.14 in [4] imply $\mathbb{E}\left[\exp\left\{\frac{a^2}{2}\mathscr{B}(T)J_T\right\}\right] < \infty$, from which item (c) follows by (4).

2. We have only to show the square integrability of $Y^{a,b}$. To this end, we firstly obtain an upper estimate of $(Y_T^{a,b})^2$, and confirm its integrability.
   Denoting $\gamma := 2b + a^2\mathscr{B}(T)$, we have

$$
\begin{aligned}
(Y_T^{a,b})^2 &= \exp\left\{ -a^2\int_0^T \sigma_s^2 ds + 2a\int_0^T \sigma_t dW_t \right. \\
&\qquad \left. + \int_0^T\int_0^\infty 2bx\widetilde{N}(dx,dt) + \int_0^T\int_0^\infty 2[bx+1-e^{bx}]\nu(dx)dt \right\} \\
&\le \exp\left\{ -2a^2\int_0^T \sigma_s^2 ds + 2a\int_0^T \sigma_t dW_t + a^2\sigma_0^2\mathscr{B}(T) + a^2\mathscr{B}(T)J_T \right. \\
&\qquad \left. + \int_0^T\int_0^\infty 2bx\widetilde{N}(dx,dt) + \int_0^T\int_0^\infty 2[bx+1-e^{bx}]\nu(dx)dt \right\} \\
&= \exp\left\{ -2a^2\int_0^T \sigma_s^2 ds + 2a\int_0^T \sigma_t dW_t + \int_0^T\int_0^\infty \gamma x\widetilde{N}(dx,dt) \right. \\
&\qquad \left. + \int_0^T\int_0^\infty \left[\gamma x + 2 - 2e^{bx}\right]\nu(dx)dt + a^2\sigma_0^2\mathscr{B}(T) \right\} \\
&= \exp\left\{ \int_0^T\int_0^\infty \left[1 - 2e^{bx} + e^{\gamma x}\right]\nu(dx)dt + a^2\sigma_0^2\mathscr{B}(T) \right\} Y_T^{2a,\gamma}.
\end{aligned}
$$

   Under (8), we have $\int_1^\infty \exp\{2\gamma x\}\nu(dx) < \infty$. Thus, we can see that $Y^{2a,\gamma}$ is a martingale by the same sort argument as item 1. Moreover, we have $\int_0^\infty \left[1 - 2e^{bx} + e^{\gamma x}\right]\nu(dx) < \infty$, from which the square integrability of $Y_T^{a,b}$ follows. □

Next, we give a definition of an LRM strategy based on Theorem 1.6 of Schweizer [13].

Definition 2.6.

1. Θ_S denotes the space of all $\mathbb{R}$-valued predictable processes ξ satisfying $\mathbb{E}\left[\int_0^T \xi_t^2 d\langle M\rangle_t + (\int_0^T |\xi_t dA_t|)^2\right] < \infty$.
2. An L^2-strategy is given by a pair $\varphi = (\xi, \eta)$, where $\xi \in \Theta_S$ and η is an adapted process such that $V(\varphi) := \xi S + \eta$ is a right continuous process with $\mathbb{E}[V_t^2(\varphi)] < \infty$ for every $t \in [0, T]$. Note that ξ_t (resp. η_t) represents the amount of units of the risky asset (resp. the risk-free asset) an investor holds at time t.
3. For claim $F \in L^2(\mathbb{P})$, the process $C^F(\varphi)$ defined by $C_t^F(\varphi) := F1_{\{t=T\}} + V_t(\varphi) - \int_0^t \xi_s dS_s$ is called the cost process of $\varphi = (\xi, \eta)$ for F.
4. An L^2-strategy φ is said a locally risk-minimizing strategy for claim F if $V_T(\varphi) = 0$ and $C^F(\varphi)$ is a martingale orthogonal to M, that is, $[C^F(\varphi), M]$ is a uniformly integrable martingale.
5. An $F \in L^2(\mathbb{P})$ admits a Föllmer-Schweizer decomposition (FS decomposition, for short) if it can be described by

$$F = F_0 + \int_0^T \xi_t^F dS_t + L_T^F, \tag{9}$$

where $F_0 \in \mathbb{R}$, $\xi^F \in \Theta_S$ and L^F is a square-integrable martingale orthogonal to M with $L_0^F = 0$.

For more details on LRM strategies, see Schweizer [12, 13]. Now, we introduce a relationship between an LRM strategy and an FS decomposition.

Proposition 2.7. *Under Assumption 2.1, an LRM strategy $\varphi = (\xi, \eta)$ for F exists if and only if F admits an FS decomposition; and its relationship is given by*

$$\xi_t = \xi_t^F, \quad \eta_t = F_0 + \int_0^t \xi_s^F dS_s + L_t^F - F1_{\{t=T\}} - \xi_t^F S_t.$$

Proof. This is from Proposition 5.2 of [13], together with Proposition 2.3. □

Thus, it suffices to get a representation of ξ^F in (9) in order to obtain an LRM strategy for claim F. Henceforth, we identify ξ^F with an LRM strategy for F.

2.2 *Minimal Martingale Measure*

We need to study the MMM in order to discuss FS decomposition. A probability measure $\mathbb{P}^* \sim \mathbb{P}$ is called the minimal martingale measure (MMM), if S is a $\mathbb{P}^*$-martingale; and any square-integrable $\mathbb{P}$-martingale orthogonal to M remains a martingale under $\mathbb{P}^*$. Now, we consider the following SDE:

$$dZ_t = -Z_{t-}\Lambda_t dM_t, \quad Z_0 = 1. \tag{10}$$

The solution to (10) is a stochastic exponential of $-\int_0^\cdot \Lambda_t dM_t$. More precisely, denoting

$$u_t := \Lambda_t S_{t-}\sigma_t = \frac{\mu}{\sigma_t} + \hat{\beta}\sigma_t \tag{11}$$

for $t \in [0, T]$, we have $\Lambda_t dM_t = u_t dW_t$; and

$$Z_t = \exp\left\{ -\frac{1}{2}\int_0^t u_s^2 ds - \int_0^t u_s dW_s \right\}. \tag{12}$$

To see that Z_T becomes the density of the MMM, it is enough to show that Z is a square integrable martingale.

Proposition 2.8. *Z is a square integrable martingale.*

Proof. It is enough to see the square integrability of Z_T, and confirm if Z satisfies Novikov's criterion from the view of Theorem III.41 in Protter [9].

First, we see $Z_T \in L^2(\mathbb{P})$. By (6), there is a constant $C_u > 0$ such that

$$u_t^2 = \frac{\mu^2}{\sigma_t^2} + 2\mu\hat{\beta} + \hat{\beta}^2\sigma_t^2 \leq C_u + \hat{\beta}^2\sigma_t^2$$

by (11). Thus, (12) implies

$$\begin{aligned}
Z_T^2 &= \exp\left\{-2\int_0^T u_t^2 dt - \int_0^T 2u_t dW_t + \int_0^T u_t^2 dt\right\} \\
&\leq \exp\left\{-2\int_0^T u_t^2 dt - \int_0^T 2u_t dW_t + TC_u + \hat{\beta}^2\int_0^T \sigma_t^2 dt\right\} \\
&\leq \exp\left\{-2\int_0^T u_t^2 dt - \int_0^T 2u_t dW_t + TC_u + \hat{\beta}^2[\sigma_0^2\mathscr{B}(T) + \mathscr{B}(T)J_T]\right\} \\
&\leq \exp\left\{TC_u + \hat{\beta}^2\sigma_0^2\mathscr{B}(T) + \int_0^T\int_0^\infty [e^{\kappa x} - 1]\,\nu(dx)dt\right\} \\
&\quad \times \exp\left\{ -2\int_0^T u_t^2 dt - \int_0^T 2u_t dW_t + \int_0^T\int_0^\infty \kappa x\widetilde{N}(dx, dt)\right. \\
&\quad \left. + \int_0^T\int_0^\infty [\kappa x + 1 - e^{\kappa x}]\,\nu(dx)dt\right\},
\end{aligned}$$

since $\hat{\beta}^2\mathscr{B}(T) \leq \kappa$ by (A2). In addition, Remark 2.2 implies

$$\mathbb{E}\left[\exp\left\{2\int_0^T u_t^2 dt\right\}\right] \leq \mathbb{E}\left[\exp\left\{2TC_u + 2\hat{\beta}^2\int_0^T \sigma_t^2 dt\right\}\right]$$
$$\leq \exp\left\{2TC_u + 2\kappa\sigma_0^2\right\}\mathbb{E}\left[e^{2\kappa J_T}\right] < \infty. \quad (13)$$

Hence, we can see that Z_T^2 is integrable by the same manner as the proof of item 1 in Lemma 2.5.

Lastly, remark that (13) gives Novikov's criterion. Thus, Proposition 2.8 follows. □

Henceforth, we denote the MMM by $\mathbb{P}^*$, that is, we have $Z_T = \frac{d\mathbb{P}^*}{d\mathbb{P}}$. Note that $dW_t^{\mathbb{P}^*} := dW_t + u_t dt$ is a Brownian motion under $\mathbb{P}^*$; and $\widetilde{N}$ remains a martingale under $\mathbb{P}^*$. Remark that we can rewrite (2) and L_T as $dS_t = S_{t-}\sigma_t dW_t^{\mathbb{P}^*}$ and $L_T = \int_0^T \sigma_s dW_s^{\mathbb{P}^*} - \frac{1}{2}\int_0^T \sigma_s^2 ds$, respectively. The following two lemmas are indispensable to formulate a Malliavin calculus under $\mathbb{P}^*$.

Lemma 2.9. *$W^{\mathbb{P}^*}$ is independent of $\widetilde{N}$; and $W_t^{\mathbb{P}^*} + \int_0^t\int_0^\infty z\widetilde{N}(ds,dz)(=: X_t^*)$ is a Lévy process under $\mathbb{P}^*$.*

Proof. This is given from Theorem 3.2 in [6]. Remark that Assumptions Z1 – Z3 in [6] are their standing assumptions. Assumptions Z1 and Z2 are satisfied in our setting from Assumption 2.1. On the other hand, Assumption Z3 does not necessarily hold, but it is not needed for Theorem 3.2 in [6]. □

Lemma 2.10. *Let $\mathbb{F}^*$ be the augmented filtration generated by X^*. Then, the filtration $\mathbb{F}$ coincides with $\mathbb{F}^*$.*

Proof. Since $W^{\mathbb{P}^*}$ is $\mathbb{F}$-adapted, it is clear that $\mathbb{F} \supset \mathbb{F}^*$. Next, we have

$$W_t = W_t^{\mathbb{P}^*} - \int_0^t u_s ds = W_t^{\mathbb{P}^*} - \int_0^t \left(\frac{\mu}{\sigma_s} + \hat{\beta}\sigma_s\right) ds$$

by (11). W is then $\mathbb{F}^*$-adapted, since σ is $\mathbb{F}^*$-adapted. Thus, we have $\mathbb{F} \subset \mathbb{F}^*$. □

2.3 Malliavin Calculus Under $\mathbb{P}^*$

Here, regarding $(\Omega, \mathscr{F}, \mathbb{P}^*)$ as the underlying probability space, we formulate a Malliavin calculus for X^* under $\mathbb{P}^*$ based on Petrou [8] and Chapter 5 of Renaud [10] from the view of Lemmas 2.9 and 2.10. Although [1] adopted the canonical Lévy space framework undertaken by Solé et al. [14], we need to take a different way to define a Malliavin derivative, since the property of the canonical Lévy space is not preserved under change of measure.

First of all, we need to prepare some notation; and define iterated integrals with respect to $W^{\mathbb{P}^*}$ and $\widetilde{N}$. Denoting $U_0 := [0, T]$ and $U_1 := [0, T] \times (0, \infty)$, we define

$$Q_0(A) := \int_A dW_t^{\mathbb{P}^*} \quad \text{for any } A \in \mathscr{B}(U_0),$$

$$Q_1(A) := \int_A \widetilde{N}(dt, dx) \quad \text{for any } A \in \mathscr{B}(U_1),$$

$$\langle Q_0 \rangle := m, \quad \text{and} \quad \langle Q_1 \rangle := m \times \nu,$$

where m is the Lebesgue measure on U_0. We denote

$$G_{(j_1,\dots,j_n)} := \left\{ (u_1^{j_1}, \dots, u_n^{j_n}) \in \prod_{k=1}^n U_{j_k} : 0 < t_1 < \cdots < t_n < T \right\}$$

for $n \in \mathbb{N}$ and $(j_1, \dots, j_n) \in \{0, 1\}^n$, where $u_k^{j_k} := t_k$ if $j_k = 0$; and $:= (t_k, x)$ if $j_k = 1$ for $k = 1, \dots, n$. We define an n-fold iterated integral as follows:

$$J_n^{(j_1,\dots,j_n)}(g_n^{(j_1,\dots,j_n)}) := \int_{G_{(j_1,\dots,j_n)}} g_n^{(j_1,\dots,j_n)}(u_1^{j_1}, \dots, u_n^{j_n}) Q_{j_1}(du_1^{j_1}) \cdots Q_{j_n}(du_n^{j_n}),$$

where $g_n^{(j_1,\dots,j_n)}$ is a deterministic function in $L^2\left(G_{(j_1,\dots,j_n)}, \bigotimes_{k=1}^n \langle Q_{j_k} \rangle\right)$. Then, Theorem 1 in [8] ensures that every $L^2(\mathbb{P}^*)$ random variable F is represented as a sum of iterated integrals, that is, we can find deterministic functions $g_n^{(j_1,\dots,j_n)} \in L^2\left(G_{(j_1,\dots,j_n)}, \bigotimes_{k=1}^n \langle Q_{j_k} \rangle\right)$ for $n \in \mathbb{N}$ and $(j_1, \dots, j_n) \in \{0, 1\}^n$ such that F has the following chaos expansion:

$$F = \mathbb{E}_{\mathbb{P}^*}[F] + \sum_{n=1}^{\infty} \sum_{(j_1,\dots,j_n) \in \{0,1\}^n} J_n^{(j_1,\dots,j_n)}(g_n^{(j_1,\dots,j_n)}). \tag{14}$$

Note that the infinite series in (14) converges in $L^2(\mathbb{P}^*)$.

Now, we define $\mathbb{D}^0$ the space of Malliavin differentiable random variables; and a Malliavin derivative operator D^0. Denoting, for $1 \le k \le n$ and $t \in (0, T)$,

$$\begin{aligned} G_{(j_1,\dots,j_n)}^k(t) := \Big\{ &(u_1^{j_1}, \dots, u_{k-1}^{j_{k-1}}, u_{k+1}^{j_{k+1}} \dots, u_n^{j_n}) \in G_{(j_1,\dots,j_{k-1} j_{k+1},\dots,j_n)} : \\ &0 < t_1 < \cdots < t_{k-1} < t < t_{k+1} < \cdots < t_n < T \Big\}, \end{aligned}$$

we define $\mathbb{D}^0$ as

$$\mathbb{D}^0 := \Bigg\{ F \in L^2(\mathbb{P}^*), F = \mathbb{E}_{\mathbb{P}^*}[F] + \sum_{n=1}^{\infty} \sum_{(j_1,\dots,j_n)\in\{0,1\}^n} J_n^{(j_1,\dots,j_n)}(g_n^{(j_1,\dots,j_n)}) :$$

$$\|g_1^{(0)}\|^2_{L^2(m)} + \sum_{n=2}^{\infty} \sum_{(j_1,\dots,j_n)\in\{0,1\}^n} \sum_{k=1}^{n} \mathbf{1}_{\{j_k=0\}}$$

$$\times \int_0^T \left\| g_n^{(j_1,\dots,j_{k-1},0,j_{k+1},\dots,j_n)}(\dots,t,\dots) \right\|^2_{L^2\left(G^k_{(j_1,\dots,j_n)}(t)\right)} dt < \infty \Bigg\}.$$

Moreover, for $F \in \mathbb{D}^0$ and $t \in [0,T]$, we define

$$D_t^0 F := g_1^{(0)}(t) + \sum_{n=2}^{\infty} \sum_{(j_1,\dots,j_n)\in\{0,1\}^n} \sum_{k=1}^{n} \mathbf{1}_{\{j_k=0\}}$$

$$\times J_{n-1}^{(j_1,\dots,j_{k-1},j_{k+1},\dots,j_n)} \left(g_n^{(j_1,\dots,j_{k-1},0,j_{k+1},\dots,j_n)}(\dots,t,\dots) \mathbf{1}_{G^k_{(j_1,\dots,j_n)}(t)} \right).$$

3 Main Results

We give explicit representations of LRM strategies for call and put options as our main results. As in [1], we consider firstly put options, since a Malliavin derivative for put options is given owing to its boundedness. LRM strategies for call options will be given as a corollary. If we dealt with call options firstly, then we would need to impose additional assumptions.

Before stating our main theorem, we prepare two propositions, one is a Malliavin derivative for put options; and the other is a Clark-Ocone type representation result for random variables in $\mathbb{D}^0$.

Proposition 3.1. *For $K > 0$, we have $(K - S_T)^+ \in \mathbb{D}^0$, and*

$$D_t^0 (K - S_T)^+ = -\mathbf{1}_{\{S_T < K\}} S_T \sigma_t.$$

Proof. The same result has been given in Proposition 4.1 of [1]. However, their framework of Malliavin calculus is different from ours as said at the beginning of Sect. 2.3. Thus, we give a new proof in the same way as [1].

For any $s \in [0,T]$, the chaos expansion for σ_s^2 is given as

$$\sigma_s^2 = e^{-\lambda s}\sigma_0^2 + \int_0^s \int_0^\infty e^{-\lambda(s-u)} x\nu(dx)du$$

$$+ \int_{[0,T]\times[0,\infty)} e^{-\lambda(s-u)} \mathbf{1}_{[0,s]\times(0,\infty)}(u,x) \widetilde{N}(du,dx),$$

which has no Brownian component. We have then $\sigma_s^2 \in \mathbb{D}^0$ and $D_t^0\sigma_s^2 = 0$ by the definitions of the set $\mathbb{D}^0$ and the operator D_t^0 in Sect. 2.3. Taking account of the boundedness of σ^2 from below, we can find a continuously differentiable function f with bounded derivative satisfying $f(\sigma_s^2) = \sigma_s$ for any $s \in [0, T]$. We have then $D_t^0\sigma_s = 0$ by Theorem 2 in [8]. Thus, Proposition 6 in [8] yields that

$$D_t^0 \int_0^T \sigma_s^2 ds = \int_t^T D_s^0\sigma_s^2 ds = 0$$

and

$$D_t^0 \int_0^T \sigma_s dW_s^{\mathbb{P}^*} = \sigma_t + \int_t^T D_s^0\sigma_s dW_s^{\mathbb{P}^*} = \sigma_t.$$

As a result, we obtain $L_T \in \mathbb{D}^0$ and $D_t^0 L_T = \sigma_t$. Next, denoting

$$f_K(r) := \begin{cases} S_0 e^r, & \text{if } r \le \log(K/S_0), \\ Kr + K(1 - \log(K/S_0)), & \text{if } r > \log(K/S_0). \end{cases}$$

we have that $f_K \in C^1(\mathbb{R})$ and $0 < f_K'(r) \le K$ for any $r \in \mathbb{R}$. Thus, Theorem 2 of [8] implies that $f_K(L_T) \in \mathbb{D}^0$ and

$$D_t^0 f_K(L_T) = f_K'(L_T) D_t^0 L_T = f_K'(L_T)\sigma_t. \tag{15}$$

Since $(K - S_T)^+ = (K - f_K(L_T))^+$, we need only to see $(K - f_K(L_T))^+ \in \mathbb{D}^0$; and calculate $D_t^0(K - f_K(L_T))^+$. To this end, we take a mollifier function φ which is a C^∞-function from $\mathbb{R}$ to $[0, \infty)$ with $\mathrm{supp}(\varphi) \subset [-1, 1]$ and $\int_{-\infty}^{\infty} \varphi(x)dx = 1$. We denote $\varphi_n(x) := n\varphi(nx)$ and $g_n(x) := \int_{-\infty}^{\infty} (K - y)^+ \varphi_n(x - y)dy$ for any $n \ge 1$. Noting that

$$g_n(x) = \int_{-\infty}^{\infty} \left(K - x + \frac{y}{n}\right)^+ \varphi(y)dy = \int_{-n(K-x)}^{\infty} \left(K - x + \frac{y}{n}\right) \varphi(y)dy,$$

we have $g_n'(x) = -\int_{-n(K-x)}^{\infty} \varphi(y)dy$, so that $g_n \in C^1$ and $|g_n'| \le 1$. Thus, Theorem 2 in [8] again implies that, for any $n \ge 1$, $g_n(f_K(L_T)) \in \mathbb{D}^0$ and

$$D_t^0 g_n(f_K(L_T)) = g_n'(f_K(L_T)) D_t^0 f_K(L_T) = g_n'(f_K(L_T)) f_K'(L_T)\sigma_t \tag{16}$$

by (15). We have then

$$\sup_{n \ge 1} \|D^0 g_n(f_K(L_T))\|_{L^2(m \times \mathbb{P}^*)}^2 \le K^2 \mathbb{E}_{\mathbb{P}^*}\left[\int_0^T \sigma_t^2 dt\right] < \infty.$$

In addition, noting that

$$|g_n(x) - (K-x)^+| = \left| \int_{-1}^{1} \left\{ \left(K - x + \frac{y}{n}\right)^+ - (K-x)^+ \right\} \varphi(y)dy \right|$$
$$\leq \frac{1}{n} \int_{-1}^{1} |y|\varphi(y)dy \leq \frac{1}{n}$$

for any $x \in \mathbb{R}$, we have $\lim_{n\to\infty} \mathbb{E}[|g_n(f_K(L_T)) - (K - f_K(L_T))^+|^2] = 0$. As a result, Lemma 3.2 below implies that $(K - f_K(L_T))^+ \in \mathbb{D}^0$. Furthermore, Lemma 2 of [8] ensures the existence of a subsequence n_k such that $D^0 g_{n_k}(f_K(L_T))$ converges to $D^0(K - f_K(L_T))^+$ in the sense of $L^2(m \times \mathbb{P}^*)$. On the other hand, we have $\lim_{n\to\infty} g'_n(x) = -\mathbf{1}_{\{x<K\}} - \mathbf{1}_{\{x=K\}} \int_0^\infty \varphi(y)dy$; and $\mathbb{P}^*(f_K(L_T) = K) = 0$ by Corollary 2.3 of [6], from which $\lim_{n\to\infty} g'_n(f_K(L_T)) = -\mathbf{1}_{\{f_K(L_T)<K\}}$ a.s. follows. Consequently, by taking a further subsequence if need be, (16) provides

$$\begin{aligned} D_t^0(K - S_T)^+ &= D_t^0(K - f_K(L_T))^+ = \lim_{k\to\infty} D_t^0 g_{n_k}(f_K(L_T)) \\ &= \lim_{k\to\infty} g'_{n_k}(f_K(L_T)) f'_K(L_T)\sigma_t = -\mathbf{1}_{\{f_K(L_T)<K\}} f'_K(L_T)\sigma_t \\ &= -\mathbf{1}_{\{S_T<K\}} S_T \sigma_t, \qquad m \times \mathbb{P}^*\text{-a.s.} \end{aligned}$$

□

Lemma 3.2. *Let F be in $L^2(\mathbb{P}^*)$, and $(F_n)_{n\geq1}$ a sequence of $\mathbb{D}^0$ converging to F in $L^2(\mathbb{P}^*)$. If $\sup_{n\geq1} \|D^0 F_n\|_{L^2(m\times\mathbb{P}^*)} < \infty$, then $F \in \mathbb{D}^0$.*

Proof. This lemma is an extension of Lemma 1.2.3 of Nualart [7]. For more details, see also the proof of Lemma 5.5.5 of [10]. □

Proposition 3.3. *For $F \in \mathbb{D}^0$, we have*

$$F = \mathbb{E}_{\mathbb{P}^*}[F] + \int_0^T \mathbb{E}_{\mathbb{P}^*}[D_t^0 F | \mathscr{F}_{t-}] dW_t^{\mathbb{P}^*} + \int_0^T \int_0^\infty \psi_{t,x} \widetilde{N}(dt, dx)$$

for some predictable process $\psi \in L^2(m \times \nu \times \mathbb{P}^*)$.

Proof. Denoting by (14) the chaos expansion of F, we have

$$\begin{aligned} F =& \mathbb{E}_{\mathbb{P}^*}[F] + \sum_{n=1}^{\infty} \sum_{(j_1,\dots,j_{n-1})\in\{0,1\}^{n-1}} \\ & \left\{ J_n^{(j_1,\dots,j_{n-1},0)}(g_n^{(j_1,\dots,j_{n-1},0)}) + J_n^{(j_1,\dots,j_{n-1},1)}(g_n^{(j_1,\dots,j_{n-1},1)}) \right\} \\ =& \mathbb{E}_{\mathbb{P}^*}[F] + \int_0^T g_1^{(0)}(t) dW_t^{\mathbb{P}^*} + \sum_{n=2}^{\infty} \sum_{(j_1,\dots,j_{n-1})\in\{0,1\}^{n-1}} \int_0^T J_{n-1}^{(j_1,\dots,j_{n-1})} \end{aligned}$$

$$\left(g_n^{(j_1,\ldots,j_{n-1},0)}(\ldots,t)\mathbf{1}_{G^n_{(j_1,\ldots,j_n)}(t)}\right) dW_t^{\mathbb{P}^*} + \int_0^T \int_0^\infty g_1^{(1)}((t,x))\widetilde{N}(dt,dx)$$

$$+\sum_{n=2}^{\infty} \sum_{(j_1,\ldots,j_{n-1})\in\{0,1\}^{n-1}} \int_0^T \int_0^\infty J_{n-1}^{(j_1,\ldots,j_{n-1})}$$

$$\left(g_n^{(j_1,\ldots,j_{n-1},1)}(\ldots,(t,x))\mathbf{1}_{G^n_{(j_1,\ldots,j_n)}(t)}\right)\widetilde{N}(dt,dx)$$

$$=\mathbb{E}_{\mathbb{P}^*}[F] + \int_0^T \left\{ g_1^{(0)}(t) + \sum_{n=2}^{\infty} \sum_{(j_1,\ldots,j_{n-1})\in\{0,1\}^{n-1}} J_{n-1}^{(j_1,\ldots,j_{n-1})}\right.$$

$$\left.\left(g_n^{(j_1,\ldots,j_{n-1},0)}(\ldots,t)\mathbf{1}_{G^n_{(j_1,\ldots,j_n)}(t)}\right)\right\} dW_t^{\mathbb{P}^*}$$

$$+\int_0^T \int_0^\infty \left\{ g_1^{(1)}((t,x)) + \sum_{n=2}^{\infty} \sum_{(j_1,\ldots,j_{n-1})\in\{0,1\}^{n-1}} J_{n-1}^{(j_1,\ldots,j_{n-1})}\right.$$

$$\left.\left(g_n^{(J_1,\ldots,j_{n-1},1)}(\ldots,(t,x))\mathbf{1}_{G^n_{(j_1,\ldots,j_n)}(t)}\right)\right\}\widetilde{N}(dt,dx) \tag{17}$$

$$=:\mathbb{E}_{\mathbb{P}^*}[F] + \int_0^T \phi_t dW_t^{\mathbb{P}^*} + \int_0^T \int_0^\infty \psi_{t,x}\widetilde{N}(dt,dx).$$

The above third equality (17) is proved in Lemma 3.4 below. On the other hand, noting that $F \in \mathbb{D}^0$, we have

$$\mathbb{E}_{\mathbb{P}^*}[D_t^0 F|\mathscr{F}_{t-}]$$

$$= \mathbb{E}_{\mathbb{P}^*}\left[g_1^{(0)}(t) + \sum_{n=2}^{\infty} \sum_{(j_1,\ldots,j_n)\in\{0,1\}^n} \sum_{k=1}^{n} \mathbf{1}_{\{j_k=0\}}\right.$$

$$\left.\times J_{n-1}^{(j_1,\ldots,j_{k-1}\,j_{k+1},\ldots,j_n)}\left(g_n^{(j_1,\ldots,j_{k-1},0,j_{k+1},\ldots,j_n)}(\ldots,t,\ldots)\mathbf{1}_{G^k_{(j_1,\ldots,j_n)}(t)}\right)\middle|\mathscr{F}_{t-}\right]$$

$$= g_1^{(0)}(t) + \sum_{n=2}^{\infty} \sum_{(j_1,\ldots,j_n)\in\{0,1\}^n} \sum_{k=1}^{n} \mathbf{1}_{\{j_k=0\}}$$

$$\times \mathbb{E}_{\mathbb{P}^*}\left[J_{n-1}^{(j_1,\ldots,j_{k-1}\,j_{k+1},\ldots,j_n)}\left(g_n^{(j_1,\ldots,j_{k-1},0,j_{k+1},\ldots,j_n)}(\ldots,t,\ldots)\mathbf{1}_{G^k_{(j_1,\ldots,j_n)}(t)}\right)\middle|\mathscr{F}_{t-}\right]$$

$$= g_1^{(0)}(t) + \sum_{n=2}^{\infty} \sum_{(j_1,\ldots,j_n)\in\{0,1\}^n} \mathbf{1}_{\{j_n=0\}} J_{n-1}^{(j_1,\ldots,j_{n-1})}\left(g_n^{(j_1,\ldots,j_{n-1},0)}(\ldots,t)\mathbf{1}_{G^n_{(j_1,\ldots,j_n)}(t)}\right)$$

$$= \phi_t.$$

As a result, ϕ belongs to $L^2(m\times\mathbb{P}^*)$. Thus, $\int_0^T\int_0^\infty \psi_{t,x}\widetilde{N}(dt,dx)$ is square integrable, that is, $\psi\in L^2(m\times\nu\times\mathbb{P}^*)$. This completes the proof of Proposition 3.3. □

Lemma 3.4. *(17) in the proof of Proposition 17 holds true. In other words, we have, for $l=0,1$,*

$$\sum_{n=2}^{\infty}\sum_{(j_1,\ldots,j_{n-1})\in\{0,1\}^{n-1}}\int_{U_l} J_{n-1}^{(j_1,\ldots,j_{n-1})}\left(g_n^{(j_1,\ldots,j_{n-1},l)}(\ldots,\hat{u}^l)\mathbf{1}_{G^n_{(j_1,\ldots,j_n)}(t)}\right)Q_l(d\hat{u}^l)$$

$$=\int_{U_l}\sum_{n=2}^{\infty}\sum_{(j_1,\ldots,j_{n-1})\in\{0,1\}^{n-1}} J_{n-1}^{(j_1,\ldots,j_{n-1})}\left(g_n^{(j_1,\ldots,j_{n-1},l)}(\ldots,\hat{u}^l)\mathbf{1}_{G^n_{(j_1,\ldots,j_n)}(t)}\right)Q_l(d\hat{u}^l),$$

where $\hat{u}^0=t\in U_0$ and $\hat{u}^1=(t,x)\in U_1$.

Proof. Recall that the infinite series in a chaos expansion converges in the $L^2(\mathbb{P}^*)$-sense. Now, for $l=0,1$, we denote

$$\Phi^{l,N}(\hat{u}^l):=\sum_{n=2}^{N}\sum_{(j_1,\ldots,j_{n-1})\in\{0,1\}^{n-1}} J_{n-1}^{(j_1,\ldots,j_{n-1})}\left(g_n^{(j_1,\ldots,j_{n-1},l)}(\ldots,\hat{u}^l)\mathbf{1}_{G^n_{(j_1,\ldots,j_n)}(t)}\right)$$

for $N\geq 2$, and

$$\Phi^{l}(\hat{u}^l):=\sum_{n=2}^{\infty}\sum_{(j_1,\ldots,j_{n-1})\in\{0,1\}^{n-1}} J_{n-1}^{(j_1,\ldots,j_{n-1})}\left(g_n^{(j_1,\ldots,j_{n-1},l)}(\ldots,\hat{u}^l)\mathbf{1}_{G^n_{(j_1,\ldots,j_n)}(t)}\right).$$

We have then that, for $l=0,1$, $(\Phi^{l,N})_{N\geq 2}$ is a sequence of $L^2(\langle Q_l\rangle\times\mathbb{P}^*)$ converging to Φ^l in the $L^2(\langle Q_l\rangle\times\mathbb{P}^*)$-sense. Thus, we have

$$\lim_{N\to\infty}\mathbb{E}_{\mathbb{P}^*}\left[\left|\int_{U_l}\Phi^{l,N}(\hat{u}^l)Q_l(d\hat{u}^l)-\int_{U_l}\Phi^{l}(\hat{u}^l)Q_l(d\hat{u}^l)\right|^2\right]=0.$$

□

The following theorem is our main result.

Theorem 3.5. *For $K>0$, an LRM strategy $\xi^{(K-S_T)^+}$ of put option $(K-S_T)^+$ is represented as*

$$\xi_t^{(K-S_T)^+}=\frac{-1}{S_{t-}}\mathbb{E}_{\mathbb{P}^*}[\mathbf{1}_{\{S_T<K\}}S_T|\mathscr{F}_{t-}]. \tag{18}$$

Proof. Denoting by ζ_t the right hand side of (18), we shall see that the process ζ is in Θ_S. Noting that $|\zeta_t| \le \frac{K}{S_{t-}}$, we have

$$\mathbb{E}\left[\int_0^T \zeta_t^2 d\langle M\rangle_t + \left(\int_0^T |\zeta_t dA_t|\right)^2\right]$$
$$\le \mathbb{E}\left[\int_0^T K^2\sigma_t^2 dt + \left(\int_0^T K\left|\mu + \left(\beta + \frac{1}{2}\right)\sigma_t^2\right| dt\right)^2\right] < \infty,$$

since $\mathbb{E}\left[\left(\int_0^T \sigma_t^2 dt\right)^2\right] < \infty$ by Lemma 2.4. As a result, $\zeta \in \Theta_S$ holds.

Next, defining

$$L_t^{(K-S_T)^+} := \mathbb{E}\left[(K-S_T)^+ - \mathbb{E}_{\mathbb{P}^*}\left[(K-S_T)^+\right] - \int_0^T \zeta_s dS_s \middle| \mathscr{F}_t\right],$$

we show that

$$(K-S_T)^+ = \mathbb{E}_{\mathbb{P}^*}\left[(K-S_T)^+\right] + \int_0^T \zeta_t dS_t + L_T^{(K-S_T)^+}$$

gives an FS decomposition of $(K-S_T)^+$. Since $L^{(K-S_T)^+}$ is a $\mathbb{P}$-martingale with $L_T^{(K-S_T)^+} \in L^2(\mathbb{P})$, we have only to show the orthogonality of $L^{(K-S_T)^+}$ to M. Since $(K-S_T)^+ \in \mathbb{D}^0$ from Proposition 3.1, we have, by Propositions 3.3 and 3.1,

$$\begin{aligned}(K-S_T)^+ &= \mathbb{E}_{\mathbb{P}^*}\left[(K-S_T)^+\right] + \int_0^T \mathbb{E}_{\mathbb{P}^*}\left[D_t^0(K-S_T)^+|\mathscr{F}_{t-}\right] dW_t^{\mathbb{P}^*} \\ &\quad + \int_0^T\int_0^\infty \psi_{t,x}\widetilde{N}(dt,dx) \\ &= \mathbb{E}_{\mathbb{P}^*}\left[(K-S_T)^+\right] - \int_0^T \mathbb{E}_{\mathbb{P}^*}\left[\mathbf{1}_{\{S_T<K\}}S_T|\mathscr{F}_{t-}\right]\sigma_t dW_t^{\mathbb{P}^*} \\ &\quad + \int_0^T\int_0^\infty \psi_{t,x}\widetilde{N}(dt,dx) \\ &= \mathbb{E}_{\mathbb{P}^*}\left[(K-S_T)^+\right] + \int_0^T \zeta_t dS_t + \int_0^T\int_0^\infty \psi_{t,x}\widetilde{N}(dt,dx)\end{aligned}$$

for some predictable process $\psi \in L^2(m \times \nu \times \mathbb{P}^*)$, which means $L_t^{(K-S_T)^+} = \int_0^t\int_0^\infty \psi_{s,x}\widetilde{N}(ds,dx)$ for any $t \in [0,T]$. Thus, the continuity of M implies that $[M, L^{(K-S_T)^+}] = 0$, which means the orthogonality of $L^{(K-S_T)^+}$ to M. □

By the put-call parity, the following holds:

Corollary 3.6. *An LRM strategy for call option* $(S_T - K)^+$ *is given as* $\xi^{(S_T-K)^+} = 1 + \xi^{(K-S_T)^+}$.

4 Conclusions

We give representations of LRM strategies of call and put options for BNS models with constraint $\rho = 0$. Compared with [1], we relax the restriction on β; and restrict ρ to 0 instead.

Now, we discuss the difference from the result in [1]. (3.1) in Theorem 3.1 of [1], giving an expression of $\xi_t^{(K-S_T)^+}$, consists of two terms. The first term is almost same as our main result given in (18) of Theorem 3.5. Note that β (or $\hat{\beta}$) does not appear in (18), although the density of the MMM is depending on β. On the other hand, the second term of (3.1) in [1], given as an integration with respect to the Lévy measure ν, seems to be corresponding to risk caused by the possible jump of S at time t. Since S is continuous in our setting, the second term is vanished in (18). As a result, (18) is given by a simple form, which coincides with (3.1) in [1] by substituting 0 for ρ.

Some important problems related to LRM strategies for BNS models still remain for future research: development of numerical scheme, comparison with delta hedge, extensions to the fully general case of BNS models, and so forth. In addition, it is indispensable to consider a simple version of the proof of Theorem 3.5, since the present proof is rather complicated although the result is simple. Furthermore, in order to understand better the results in this paper, we need to extend our results to more general stochastic volatility models.

Acknowledgements The author would like to thank to Jean-Pierre Fouque for fruitful discussion, and an anonymous referee for valuable comments and suggestions. This research was supported by Ishii memorial securities research promotion foundation.

References

1. Arai T, Suzuki R (2016) Local risk minimization for Barndorff-Nielsen and Shephard models. submitted. Available at http://arxiv.org/pdf/1503.08589v2
2. Barndorff-Nielsen OE, Shephard N (2001) Modelling by Lévy processes for financial econometrics. In: Barndorff-Nielsen OE, Mikosch T, Resnick S (eds) Lévy processes – theory and applications. Birkhäuser, Basel, pp 283–318
3. Barndorff-Nielsen OE, Shephard N (2001) Non-Gaussian Ornstein-Uhlenbeck based models and some of their uses in financial econometrics. J R Stat Soc 63:167–241
4. Cont R, Tankov P (2004) Financial modelling with jump processes. Chapman & Hall, London
5. Ishikawa Y (2013) Stochastic calculus of variations for jump processes. Walter De Gruyter, Berlin

6. Nicolato E, Venardos E (2003) Option pricing in stochastic volatility models of the Ornstein-Uhlenbeck type. Math Financ 13(4):445–466
7. Nualart D (1995) The Malliavin calculus and related topics. Springer, Berlin/New York
8. Petrou E (2008) Malliavin calculus in Lévy spaces and applications to finance. Electron J Probab 27:852–879
9. Protter P (2004) Stochastic integration and differential equations. Springer, Berlin
10. Renaud JF (2007) Calcul de Malliavin, processus de Lévy et applications en finance: quelques contributions. Dissertation, Université de Montréal. Available at http://neumann.hec.ca/pages/bruno.remillard/Theses/JFRenaud.pdf
11. Schoutens W (2003) Lévy processes in finance: pricing financial derivatives. Wiley, Hoboken
12. Schweizer M (2001) A guided tour through quadratic hedging approaches. In: Jouini E, Cvitanić J, Musiela M (eds) Option pricing, interest rates and risk management. Handbooks in mathematical finance. Cambridge University Press, Cambridge, pp 538–574
13. Schweizer M (2008) Local risk-minimization for multidimensional assets and payment streams. Banach Center Publ 83:213–229
14. Solé JL, Utzet F, Vives J (2007) Canonical Lévy process and Malliavin calculus. Stoch Process Appl 117:165–187
15. Wang W, Qian L, Wang W (2013) Hedging strategy for unit-linked life insurance contracts in stochastic volatility models. WSEAS Trans Math 12(4):363–373

Adv. Math. Econ. 20, 23–75 (2016)

Advances in
MATHEMATICAL
ECONOMICS

On a Fractional Differential Inclusion in Banach Space Under Weak Compactness Condition

C. Castaing, C. Godet-Thobie, L.X. Truong, and F.Z. Mostefai

Abstract We consider a class of boundary value problem in a separable Banach space governed by a fractional differential inclusion with integral boundary conditions

$$\begin{cases} w\text{-}D^{\alpha}u(t) \in F(t, u(t), w\text{-}D^{\alpha-1}u(t)), \ \ t \in [0, 1] \\ I^{\beta}u(t)|_{t=0} = 0, \ \ u(1) = \int_0^1 u(t)dt \end{cases}$$

JEL Classification: C61, C73

Mathematics Subject Classification (2010): 34A60, 34B15, 47H10

C. Castaing (✉)
Département de Mathématiques, Université Montpellier II, Case Courrier 051, 34095 Montpellier Cedex 5, France
e-mail: charles.castaing@gmail.com

C. Godet-Thobie
Laboratoire de Mathématiques de Bretagne Atlantique, Université de Bretagne Occidentale, CNRS UMR 6205, 6, avenue Victor Le Gorgeu, CS 9387, F-29238 Brest Cedex 3, France
e-mail: christiane.godet-thobie@univ-brest.fr

L.X. Truong
Department of Mathematics and Statistics, University of Economics of HoChiMinh City, 59C Nguyen Dinh Chieu Str. Dist. 3 HoChiMinh City, Vietnam
e-mail: lxuantruong@gmail.com

F.Z. Mostefai
Département de Mathématiques, Université de Saida, BP 138, Cite Ennasr, 20000 Saida, Algérie
e-mail: fatymath@gmail.com

S. Kusuoka, T. Maruyama (eds.), *Advances in Mathematical Economics Volume 20*, Advances in Mathematical Economics,
DOI 10.1007/978-981-10-0476-6_2

where $\alpha \in]1,2]$, $\beta \in]0,\infty[$ are given constant and $w\text{-}D^{\gamma}$ is the fractional w-R.L derivative of order $\gamma \in \{\alpha - 1, \alpha\}$, F is a convex weakly compact valued mapping. Topological properties of the solutions set are presented. Applications to control problems and further variants are provided.

Keywords Fractional differential inclusion • w-R.L derivative • Pettis integral • Relaxation • Young measures

Article Type: Research Article
Received: November 25, 2015
Revised: December 6, 2015

1 Introduction and Preliminaries

Fractional differential equations have recently proved to be valuable tools in the modeling of many phenomena in various fields of science and engineering. There are numerous applications to viscoelasticity, control, economics and so forth. Subsequently, there has been a great deal of research on this field, see for examples, [1–3, 9, 10, 22, 25–27, 30–33, 35–37] and the references therein. The present work is a contribution in this direction.

In this paper, we investigate a class of boundary value problems governed by a fractional differential inclusion (FDI) in a separable Banach space E in both Bochner and Pettis settings

$$\begin{cases} w\text{-}D^{\alpha}u(t) \in F(t, u(t), w\text{-}D^{\alpha-1}u(t)), & t \in [0,1] \\ I^{\beta}u(t)|_{t=0} = 0, \ u(1) = \int_0^1 u(t)dt & \end{cases} \tag{1}$$

where $\alpha \in]1,2]$, $\beta \in]0,\infty[$ are given constants, $w\text{-}D^{\gamma}$ is the fractional w-R.L derivative of order $\gamma \in \{\alpha - 1, \alpha\}$, $F : [0,1] \times E_{\sigma} \times E_{\sigma} \hookrightarrow E_{\sigma}$ is a convex weakly compact valued multimapping, E_{σ} is the vector space E endowed with the weak topology and $F(t,.,.) : E_{\sigma} \times E_{\sigma} \hookrightarrow E_{\sigma}$ is upper semicontinuous. We refer to [7, 14–16, 23] dealing with ordinary differential equation (ODE) or second order ordinary differential equation (SODE) in which several related problems are also investigated.

In the present study, we provide several new results in fractional differential inclusion (FDI), with applications to Bolza and Relaxation problem in Optimal Control with Young measure using new tools and subtile techniques. This paper is organized as follows. In Sect. 2, we introduce some notions and recall some definitions and needed results. In particular, we provide some results for the fractional calculus in the Bochner and Pettis contexts. In Sect. 3, we provide existence results of $W^{\alpha,1}_{B,E}([0,1])$-solutions and topological properties of the solutions set to (1) in the Bochner setting. Section 4 is devoted to $W^{\alpha,1}_{P,E}([0,1])$-solutions and topological properties of the solutions set to (1) in the Pettis setting, here several variants are provided. Section 5 is devoted to the fractional integral inclusion with applications. In Sect. 6, we present an application to relaxation problem in optimal control governed by fractional differential equation (FDE) involving Young measure [8, 18, 28].

The stated results are new in an active field of research and lead to further developments in second order differential inclusion with boundary condition, viscosity and variational convergence [6, 14, 19], fractional integral inclusion and also fractional integration of fuzzy level sets.

2 Preliminaries on the Fractional Integral

On the fractional Bochner integral

For the sake of completeness, we state and summarize at first in this section some needed properties for the fractional calculus in Bochner case. Let $\lambda := dt$ be the Lebesgue measure on $[0, 1]$ and let $\mathcal{L}_\lambda([0, 1])$ the σ-algebra of Lebesgue measurable sets on $[0, 1]$. E is a separable Banach space, $\overline{B}_{E^*}$ is the closed unit ball in the dual E^* and $L^1_E([0, 1])$ is the space of all Bochner integrable mappings $f : [0, 1] \to E$.

We provide some definitions from fractional calculus which will be used in the sequel. Throughout we assume $\alpha \in]1, 2]$ and $\beta \in]0, \infty[$.

Definition 2.1. Let $u : [0, 1] \to E$. The fractional w-R.L derivative of order $\gamma > 0$ of the function u, $w\text{-}D^\gamma u : [0, 1] \to E$, is defined by

$$\langle x^*, w\text{-}D^\gamma u(t)\rangle = D^\gamma \langle x^*, u(t)\rangle, \forall t \in [0, 1], \forall x^* \in E^*$$

where $D^\gamma \langle x, u(t)\rangle$ denotes the Riemann-Liouville fractional derivative of order γ of the real valued function $t \mapsto \langle x, u(t)\rangle$.

The following lemma is useful for our purpose.

Lemma 2.1. *Let $f \in L^1_E([0, 1])$. Then $I^\alpha f : [0, 1] \to E$ is continuous, that is $I^\alpha f \in C_E([0, 1])$.*

Proof. Lemma 2.1 follows from Lemma 2.3 below. ∎

Denote by

$$W^{\alpha,1}_{B,E}([0, 1]) = \{f \in C_E([0, 1]) : w\text{-}D^{\alpha-1}f \in C_E([0, 1]);\ w\text{-}D^\alpha f \in L^1_E[0, 1]\}$$

where $w\text{-}D^{\alpha-1}f$ and $w\text{-}D^\alpha f$ are the fractional w-R.L derivative of order $\alpha - 1$ and α of f, respectively.

Definition 2.2. A function $u \in W^{\alpha,1}_{B,E}([0, 1])$ is a solution to problem (1) if there exists a function $g \in L^1_E([0, 1])$ such that

$$\begin{cases} w\text{-}D^\alpha u(t) = g(t) \in F(t, u(t), w\text{-}D^{\alpha-1}u(t)), & t \in [0, 1] \\ I^\beta u(t)|_{t=0} = 0,\ u(1) = \int_0^1 u(t)dt \end{cases}$$

Let $G : [0, 1] \times [0, 1] \to \mathbf{R}$ be defined by

$$G(t,s) = \begin{cases} \frac{(t-s)^{\alpha-1}}{\Gamma(\alpha)}, & 0 \le s \le t \le 1 \\ 0, & 0 \le t < s \le 1 \end{cases} + \frac{t^{\alpha-1}}{(\alpha-1)\,\Gamma(\alpha)} \left((1-s)^{\alpha} - \alpha\,(1-s)^{\alpha-1} \right). \tag{2}$$

It is worth to mention that G is Borel, $G(.,s)$ is continuous for all $s \in [0, 1]$ and

$$|G(t,s)| \le \frac{2\alpha}{(\alpha-1)\Gamma(\alpha)} := M_G$$

for all $t, s \in [0, 1]$.

Lemma 2.2. *Let G as above. Let* $f \in L^1_E([0,1])$ *and let* $u_f(t) = \int_0^1 G(t,s)f(s)ds$, $t \in [0,1]$. *Then*

(i) $I^{\beta}u_f(t)|_{t=0} = 0$, $u_f(1) = \int_0^1 u_f(t)dt$,
(ii) $u_f \in W^{\alpha,1}_{B,E}([0,1])$,
(iii) $w\text{-}D^{\alpha}u_f(t) = f(t)$ *a.e.,*
(iv) $w\text{-}D^{\alpha-1}u_f(t) = \int_0^t f(s)ds + \frac{1}{\alpha-1}\int_0^1 ((1-s)^{\alpha} - \alpha(1-s)^{\alpha-1})f(s)ds$, $t \in [0,1]$.

Proof. Let $x^* \in E^*$. For all $t \in [0, 1]$, we have

$$\langle x^*, I^{\beta}u_f(t)\rangle = I^{\beta}\langle x^*, u_f(t)\rangle = I^{\beta}[\int_0^1 G(t,s)\langle x^*, f(s)\rangle ds]$$
$$= I^{\alpha+\beta}\langle x^*, f(t)\rangle + I^{\beta}[\frac{\alpha t^{\alpha-1}}{\alpha-1}\langle x^*, I^{\alpha+1}f(1) - I^{\alpha}f(1)\rangle]$$
$$= I^{\alpha+\beta}\langle x^*, f(t)\rangle + \frac{\alpha\Gamma(\alpha)}{(\alpha-1)\Gamma(\alpha+\beta)}\langle x^*, I^{\alpha+1}f(1) - I^{\alpha}f(1)\rangle t^{\alpha+\beta-1}.$$

Letting $t \to 0^+$ we get

$$\lim_{t\to 0^+}\langle x^*, I^{\beta}u_f(t)\rangle = \lim_{t\to 0^+} I^{\alpha+\beta}\langle x^*, f(t)\rangle = 0. \tag{3}$$

By definition (2) of G we have

$$u_f(t) = I^{\alpha}f(t) + \frac{\alpha t^{\alpha-1}}{\alpha-1}[I^{\alpha+1}f(1) - I^{\alpha}f(1)], \quad t \in [0,1]. \tag{4}$$

By Lemma 2.1, $I^{\alpha}f$ is continuous, so is u_f. From (4) it follows that

$$u_f(1) = \frac{1}{\alpha-1}[\alpha I^{\alpha+1}f(1) - I^{\alpha}f(1)] \tag{5}$$

and

$$\int_0^1 u_f(t)dt = \int_0^1 I^\alpha f(t)dt + \frac{1}{\alpha - 1}[I^{\alpha+1}f(1) - I^\alpha f(1)] \tag{6}$$
$$= I^{\alpha+1}f(1) + \frac{1}{\alpha - 1}[I^{\alpha+1}f(1) - I^\alpha f(1)]$$
$$= \frac{1}{\alpha - 1}[\alpha I^{\alpha+1}f(1) - I^\alpha f(1)].$$

From (5) and (6) we get

$$\int_0^1 u_f(t)dt = u_f(1). \tag{7}$$

On the other hand, we have

$$\langle x^*, w\text{-}D^\alpha u_f(t)\rangle = D^\alpha \langle x^*, u_f(t)\rangle = \langle x^*, f(t)\rangle, \quad t \in [0, 1] \tag{8}$$

and

$$\langle x^*, w\text{-}D^{\alpha-1}u_f(t)\rangle = D^{\alpha-1}\left\langle x^*, u_f(t)\right\rangle$$
$$= D^{\alpha-1}\left(I^\alpha \langle x^*, f(t)\rangle\right)$$
$$+ \left(\frac{1}{(\alpha-1)\Gamma(\alpha)}\int_0^1 \left((1-s)^\alpha - \alpha(1-s)^{\alpha-1}\right)\langle x^*, f(s)\rangle\, ds\right) D^{\alpha-1}\left(t^{\alpha-1}\right)$$
$$= \int_0^t \langle x^*, f(s)\rangle\, ds + \frac{1}{(\alpha-1)}\int_0^1 \left((1-s)^\alpha - \alpha(1-s)^{\alpha-1}\right)\langle x^*, f(s)\rangle\, ds,$$

so that, for each $t \in [0, 1]$

$$w\text{-}D^{\alpha-1}u_f(t) = \int_0^t f(s)ds + \frac{1}{\alpha - 1}\int_0^1 ((1-s)^\alpha - \alpha(1-s)^{\alpha-1})f(s)ds. \tag{9}$$

From (9) we note that $w\text{-}D^{\alpha-1}u_f$ is continuous on $[0, 1]$. Thus $u_f \in W_{B,E}^{\alpha,1}([0, 1])$ and the proof is complete. ■

Remark. It is worth to mention that

$$(^*)\ ||u_f(t)|| = \sup_{x^*\in\overline{B}_{E^*}} |\langle x^*, \int_0^1 G(t,s)f(s)ds\rangle|$$

$$\leq \sup_{x^*\in\overline{B}_{E^*}} \int_0^1 |\langle x^*, G(t,s)f(s)\rangle|ds$$

$$\leq M_G \sup_{x^*\in\overline{B}_{E^*}} \int_0^1 |\langle x^*, f(s)\rangle|ds \leq M_G||f||_{L^1_E},\quad \forall t\in[0,1]$$

and similarly thanks to Lemma 2.2 (iv)

$$(^{**})\ ||w\text{-}D^{\alpha-1}u_f(t)|| \leq ||f||_{L^1_E} + \frac{1}{\alpha-1}||f||_{L^1_E} \leq M_G||f||_{L^1_E},\quad \forall t\in[0,1]$$

because from the properties of the Gamma function and $\alpha\in]1,2]$, we have $0 < \Gamma(\alpha) < 2$. Some arguments and notations given above will be used in the next section.

On the fractional Pettis integral

We recall and summarize at first some needed results on the Pettis integration. Let $\lambda := dt$ be the Lebesgue measure on $[0,1]$ and let $\mathcal{L}_\lambda([0,1])$ the σ-algebra of Lebesgue measurable sets on $[0,1]$. A mapping $f : [0,1] \to E$ is scalarly integrable function if , for every $x^* \in E^*$, the scalar function $\langle x^*, f\rangle$ is Lebesgue integrable on $[0,1]$. A scalarly integrable function $f : [0,1] \to E$ is Pettis-integrable if, for every Lebesgue measurable set A the weak integral $\int_A f dt$ defined by $\langle x^*, \int_A f dt\rangle = \int_A \langle x^*, f\rangle\, dt$ for all $x^* \in E^*$ belongs to E. We denote by $P^1_E([0,1])$ the space of all Pettis integrable functions $f : [0,1] \to E$ endowed with the Pettis norm $||f||_{P^1_E} = \sup_{x^*\in\overline{B}_{E^*}} \int_0^1 |\langle x^*, f\rangle| dt$. A mapping $f : [0,1] \to E$ is Pettis-integrable iff the set $\{\langle x^*, f\rangle : ||x^*|| \leq 1\}$ is uniformly integrable in the space $L^1_{\mathbf{R}}([0,1])$. More generally a convex weakly compact valued mapping $X : [0,1] \hookrightarrow E$ is scalarly integrable, if, for every $x^* \in E^*$, the scalar function $\delta^*(x^*, X)$ is integrable on $[0,1]$, X is *Pettis-integrable* if the set $\{\delta^*(x^*, X) : ||x^*|| \leq 1\}$ is uniformly integrable in the space $L^1_{\mathbf{R}}([0,1])$. In view of ([13], Theorem 4.2 or [18], Cor. 6.3.3) the set S^{Pe}_X of all Pettis-integrable selections of a convex weakly compact valued Pettis-integrable mapping $X : [0,1] \hookrightarrow E$ is sequentially $\sigma(P^1_E, L^\infty \otimes E^*)$-compact. A subset $\mathcal{H} \subset P^1_E([0,1])$ is *Pettis uniformly integrable* (PUI for short) if, for every $\varepsilon > 0$, there exists $\delta > 0$ such that

$$\lambda(A) \leq \delta \to \sup_{X\in\mathcal{H}} ||1_A X||_{P^1_E} \leq \varepsilon.$$

If $f \in P^1_E([0,1])$ the singleton $\{f\}$ is PUI since the set $\{\langle x^*, f\rangle : ||x^*|| \leq 1\}$ is uniformly integrable. More generally, a subset $\mathcal{H} \subset P^1_E([0,1])$ is *scalarly uniformly* *integrable* [4] if the set $\{\langle x^*, X\rangle : X \in \mathcal{H}, ||x^*|| \leq 1\}$ is uniformly integrable in the space $L^1_{\mathbf{R}}([0,1])$. If $\mathcal{H}$ is scalarly uniformly integrable, then it is PUI. Indeed, we have

$$\lim_{a\to\infty} \sup_{X\in\mathcal{H}} \sup_{x^*\in\overline{B}_{E^*}} \int_{|\langle x^*,X\rangle|>a} |\langle x^*, X\rangle|\, dt = 0.$$

For any $x^* \in \overline{B}_{E^*}$, one has

$$\int_A |\langle x^*, X\rangle|\, dt = \int_{A\cap[|\langle x^*,X\rangle|\leq a]} |\langle x^*, X\rangle|\, dt + \int_{A\cap[|\langle x',X\rangle|>a]} |\langle x^*, X\rangle|\, dt.$$

Let a be large enough in order to ensure

$$\forall x^* \in \overline{B}_{E^*},\ \ \forall X \in \mathcal{H},\ \int_{A\cap[|\langle x^* X\rangle|>a]} |\langle x^*, X\rangle|\, dt \leq \varepsilon/2. \qquad (*)$$

Thus, the last term of (*) is $\leq \varepsilon/2$. Now, if δ is small enough in order to ensure $a\delta \leq \varepsilon/2$, we obtain

$$\int_{A\cap[|\langle x^* X\rangle|\leq a]} |\langle x^*, X\rangle|\, dt \leq a\lambda(A) \leq \varepsilon/2$$

as soon as $\lambda(A) \leq \delta$. The preceding considerations are useful in further developments.

We provide some definitions from fractional calculus which will be used in the sequel. Throughout we assume $\alpha \in]1,2]$ and $\beta \in]0,\infty[$.

Definition 2.3. Let $f \in P^1_E([0,1])$. The fractional Pettis integral of order α of the function f is defined by

$$I^\alpha f(t) := \frac{1}{\Gamma(\alpha)} \int_0^t (t-s)^{\alpha-1} f(s)ds,\ \ t \in [0,1].$$

Now we provide a series of lemmas we need for the study of (1) in the Pettis setting.

Lemma 2.3. *Let $H : [0,1] \times [0,1] \to \mathbf{R}$ be a mapping with the following properties*

(i) for each $t \in [0,1]$, $H(t,.)$ is Lebesgue-measurable on $[0,1]$,
(ii) for each $s \in [0,1]$, $H(.,s)$ is continuous on $[0,1]$,
(iii) there is a constant $M > 0$ such that $|H(t,s)| \leq M$ for all $(t,s) \in [0,1] \times [0,1]$.

Let $f : [0, 1] \to E$ be a Pettis-integrable mapping. Then the mapping

$$u_f : t \mapsto \int_0^1 H(t, s)f(s)ds$$

is continuous from $[0, 1]$ into E, that is, $u_f \in C_E([0, 1])$.

Proof. Let (t_n) be a sequence in $[0, 1]$ such that $t_n \to t \in [0, 1]$. Then we have the estimation

$$\sup_{x^* \in \overline{B}_{E^*}} |\langle x^*, \int_0^1 H(t_n, s)f(s)ds - \int_0^1 H(t, s)f(s)ds\rangle|$$

$$\leq \sup_{x^* \in \overline{B}_{E^*}} \int_0^1 |H(t_n, s) - H(t, s)||\langle x^*, f(s)\rangle|ds.$$

As the sequence $(|H(t_n, .) - H(t, .)|)$ is bounded in $L^\infty_{\mathbf{R}}([0, 1])$ and pointwise converges to 0, it converges to 0 uniformly on uniformly integrable subsets of $L^1_{\mathbf{R}}([0, 1])$ in view of a lemma due to Grothendieck [29], in others terms it converges to 0 with respect to the Mackey topology $\tau(L^\infty, L^1)$, see also [12] for a more general result concerning the Mackey topology for bounded sequences in $L^\infty_{E^*}$. Since the set $\{|\langle x^*, f(s)\rangle| : ||x^*|| \leq 1\}$ is uniformly integrable in $L^1_{\mathbf{R}}([0, 1])$, the second term in the above estimation goes to 0 as $t_n \to t$ showing that u_f is continuous on $[0, 1]$ with respect to the strong topology of E. ■

The following lemma is crucial for our purpose. It justifies the definition and gives the continuity property for the fractional Pettis integral $I^\alpha f$.

Lemma 2.4. *Let $f \in P^1_E([0, 1])$. Then $I^\alpha f : [0, 1] \to E$ is continuous, that is $I^\alpha f \in C_E([0, 1])$.*

Proof. Lemma 2.4 is a direct consequence of Lemma 2.3 by taking $H(t, s) = \frac{(t-s)^{\alpha-1}}{\Gamma(\alpha)}$ for all $(t, s) \in [0, 1] \times [0, 1]$. ■

Denote by

$$W^{\alpha,1}_{P,E}([0, 1]) = \{f \in C_E([0, 1]) : w\text{-}D^{\alpha-1}f \in C_E([0, 1]);\ w\text{-}D^\alpha f \in P^1_E[0, 1]\}$$

where $w\text{-}D^{\alpha-1}f$ and $w\text{-}D^\alpha f$ are the fractional w-R.L derivative of order $\alpha - 1$ and α of f, respectively.

Definition 2.4. A function $u \in W^{\alpha,1}_{P,E}([0, 1])$ is a solution to problem (1) if there exists a function $g \in P^1_E([0, 1])$ such that

$$\begin{cases} w\text{-}D^\alpha u(t) = g(t) \in F(t, u(t), w\text{-}D^{\alpha-1}u(t))), \ t \in [0, 1] \\ I^\beta u(t)|_{t=0} = 0,\ u(1) = \int_0^1 u(t)dt \end{cases}$$

Lemma 2.5. *Let G be defined by (2). Let $f \in P_E^1([0,1])$ and let $u_f(t) = \int_0^1 G(t,s)f(s)ds, \quad t \in [0,1]$. Then the following hold*

(i) $I^\beta u_f(t)|_{t=0} = 0,\ u_f(1) = \int_0^1 u_f(t)dt,$
(ii) $u_f \in W_{P,E}^{\alpha,1}([0,1]),$
(iii) $w\text{-}D^\alpha u_f(t) = f(t)$ *a.e.,*
(iv) $w\text{-}D^{\alpha-1}u_f(t) = \int_0^t f(s)ds + \frac{1}{\alpha-1}\int_0^1((1-s)^\alpha - \alpha(1-s)^{\alpha-1})f(s)ds, \quad t \in [0,1].$

Proof. Let $x^* \in E^*$. We have

$$\begin{aligned}\langle x^*, I^\beta u_f(t)\rangle &= I^\beta\langle x^*, u_f(t)\rangle = I^\beta\Big[\int_0^1 G(t,s)\langle x^*, f(s)\rangle ds\Big]\\ &= I^{\alpha+\beta}\langle x^*, f(t)\rangle + I^\beta\Big[\frac{\alpha t^{\alpha-1}}{\alpha-1}\langle x^*, I^{\alpha+1}f(1) - I^\alpha f(1)\rangle\Big]\\ &= I^{\alpha+\beta}\langle x^*, f(t)\rangle + \frac{\alpha\Gamma(\alpha)}{(\alpha-1)\Gamma(\alpha+\beta)}\langle x^*, I^{\alpha+1}f(1) - I^\alpha f(1)\rangle t^{\alpha+\beta-1}.\end{aligned}$$

Letting $t \to 0^+$ we get

$$\lim_{t\to 0^+}\langle x^*, I^\beta u_f(t)\rangle = \lim_{t\to 0^+} I^{\alpha+\beta}\langle x^*, f(t)\rangle = 0.$$

By definition (2) of G, we have

$$u_f(t) = I^\alpha f(t) + \frac{\alpha t^{\alpha-1}}{\alpha-1}[I^{\alpha+1}f(1) - I^\alpha f(1)], \quad t \in [0,1]. \tag{10}$$

By Lemma 2.4, $I^\alpha f$ is continuous, so is u_f. From (10) it follows that

$$u_f(1) = \frac{1}{\alpha-1}[\alpha I^{\alpha+1}f(1) - I^\alpha f(1)] \tag{11}$$

and

$$\begin{aligned}\int_0^1 u_f(t)dt &= \int_0^1 I^\alpha f(t)dt + \frac{1}{\alpha-1}[I^{\alpha+1}f(1) - I^\alpha f(1)]\\ &= I^{\alpha+1}f(1) + \frac{1}{\alpha-1}[I^{\alpha+1}f(1) - I^\alpha f(1)]\\ &= \frac{1}{\alpha-1}[\alpha I^{\alpha+1}f(1) - I^\alpha f(1)].\end{aligned} \tag{12}$$

From (11) and (12) we get

$$\int_0^1 u_f(t)dt = u_f(1). \tag{13}$$

On the other hand, we have

$$\langle x^*, w\text{-}D^{\alpha} u_f(t)\rangle = D^{\alpha}\langle x^*, u_f(t)\rangle = \langle x^*, f(t)\rangle, \quad t \in [0,1] \tag{14}$$

and

$$\begin{aligned}\langle x^*, w\text{-}D^{\alpha-1} u_f(t)\rangle &= D^{\alpha-1}\left\langle x^*, u_f(t)\right\rangle \\ &= D^{\alpha-1}\left(I^{\alpha}\left\langle x^*, f(t)\right\rangle\right)\end{aligned}$$

$$+\left(\frac{1}{(\alpha-1)\,\Gamma(\alpha)}\int_0^1\left((1-s)^{\alpha}-\alpha(1-s)^{\alpha-1}\right)\langle x^*, f(s)\rangle\, ds\right) D^{\alpha-1}\left(t^{\alpha-1}\right)$$

$$= \int_0^t \langle x^*, f(s)\rangle\, ds + \frac{1}{(\alpha-1)}\int_0^1\left((1-s)^{\alpha}-\alpha(1-s)^{\alpha-1}\right)\langle x^*, f(s)\rangle\, ds,$$

so that, for each $t \in [0,1]$

$$w\text{-}D^{\alpha-1}u_f(t) = \int_0^t f(s)ds + \frac{1}{\alpha-1}\int_0^1((1-s)^{\alpha}-\alpha(1-s)^{\alpha-1})f(s)ds. \tag{15}$$

From (15), we note that $w\text{-}D^{\alpha-1}u_f$ is continuous on $[0,1]$ using the fact that $\{f\}$ is PUI and the argument given in Lemma 2.3. Thus $u_f \in W^{\alpha,1}_{P,E}([0,1])$ and the proof is complete. ∎

Remark. It is worth to mention that

$$\begin{aligned}(*)\ \|u_f(t)\| &= \sup_{x^*\in \overline{B}_{E^*}} |\langle x^*, \int_0^1 G(t,s)f(s)ds\rangle| \\ &\le \sup_{x^*\in \overline{B}_{E^*}} \int_0^1 |\langle x^*, G(t,s)f(s)\rangle|ds \\ &\le M_G \sup_{x^*\in \overline{B}_{E^*}} \int_0^1 |\langle x^*, f(s)\rangle|ds = M_G\|f\|_{P^1_E},\ \ \forall t\in[0,1]\end{aligned}$$

and similarly thanks to Lemma 2.5 (iv)

$$(**)\ \|w\text{-}D^{\alpha-1}u_f(t)\| \le \|f\|_{P^1_E} + \frac{1}{\alpha-1}\|f\|_{P^1_E} \le M_G\|f\|_{P^1_E},\ \ \forall t\in[0,1]$$

because from the properties of the Gamma function and $\alpha \in]1,2]$, we have $0 < \Gamma(\alpha) < 2$ so that

$$\frac{2\alpha}{2(\alpha-1)} < \frac{2\alpha}{(\alpha-1)\Gamma(\alpha)} = M_G.$$

3 Fractional Differential Inclusion in Bochner Setting

We summarize at first some needed results. Let E be a Banach space and let $\overline{B}_E$ be the closed unit ball of E.

Lemma 3.1. *Let $H \subset E$ be a subset of a Banach space E such that for each $\varepsilon > 0$, there is a weakly compact subset K such that $H \subset K + \varepsilon\overline{B}_E$. Then H is relatively weakly compact in E.*

Proof. See Grothendiek ([29], page 296). ■

Lemma 3.2. *Assume that $Q \subset C_E([0,1])$ is bounded equicontinuous such that $Q(t)$ is relatively weakly compact in E for each $t \in [0,1]$, then Q is relatively weakly compact in $C_E([0,1])$.*

The proof makes use of Lemma 3.1 and the measure of weak noncompactness [5, 24]. See e.g. the proof of Theorem 2.8 in [34].

Now we proceed to the study of fractional differential inclusion (FDI) of the form

$$\begin{cases} w\text{-}D^{\alpha}u(t) \in F(t, u(t), w\text{-}D^{\alpha-1}u(t)), & t \in [0,1] \\ I^{\beta}u(t)|_{t=0} = 0, \ u(1) = \int_0^1 u(t)dt \end{cases}$$

where $\alpha \in]1,2]$, $\beta \in]0,\infty[$ and $w\text{-}D^{\alpha}$ and $w\text{-}D^{\alpha-1}$ are the fractional w-R.L derivatives of order α and $\alpha-1$ respectively, F is a convex weakly compact valued mapping.

Here we consider at first the Bochner case. As a consequence of Lemma 2.2, we have the following result.

Lemma 3.3. *Let $f \in L^1_E([0,1])$ be a Bochner-integrable function. Then the function*

$$u_f(t) = \int_0^1 G(t,s)f(s)ds, \quad t \in [0,1]$$

is the unique $W^{\alpha,1}_{B,E}([0,1])$-solution to the fractional differential equation (FDE)

$$\begin{cases} w\text{-}D^{\alpha}u(t) = f(t), & t \in [0,1] \\ I^{\beta}u(t)|_{t=0} = 0, \ u(1) = \int_0^1 u(t)dt \end{cases}$$

This result is useful in establishing the study of the fractional differential inclusion (FDI).

Theorem 3.1. *Let $X : [0,1] \hookrightarrow E$ be a convex weakly compact valued measurable and integrably bounded mapping (i.e.$|X| \in L^1_{\mathbf{R}}([0,1])$). Then the $W^{\alpha,1}_{B,E}([0,1])$-solutions set to the (FDI)*

$$\begin{cases} w\text{-}D^{\alpha}u(t) \in X(t), \; t \in [0,1] \\ I^{\beta}u(t)|_{t=0} = 0, \; u(1) = \int_0^1 u(t)dt \end{cases} \tag{16}$$

is weakly compact in $C_E([0,1])$.

Proof. By virtue of Lemma 3.3, the $W^{\alpha,1}_{B,E}([0,1])$-solutions set $\mathcal{X}$ to (16) is characterized by

$$\mathcal{X} = \{u_f : [0,1] \to E, \; u_f(t) = \int_0^1 G(t,s)f(s)ds, \; f \in S^1_X, \; t \in [0,1]\}$$

Claim. $\mathcal{X}$ is bounded, convex, equicontinuous and weakly compact in $C_E([0,1])$.

From definition of function G in (2), it is not difficult to show that $\{u_f : f \in S^1_X\}$ is bounded, equicontinuous in $C_E([0,1])$. Indeed, let $\tau, t \in [0,1]$, we have the estimate

$$\begin{aligned} \left\|u_f(\tau) - u_f(t)\right\| &\le \int_0^1 |G(\tau,s) - G(t,s)|\,|X(s|ds \\ &\le \frac{1}{\Gamma(\alpha)} \int_0^{\tau} \left|(\tau-s)^{\alpha-1} - (t-s)^{\alpha-1}\right| |X(s)|ds \\ &\quad + \int_{\tau}^{t} \frac{(t-s)^{\alpha-1}}{\Gamma(\alpha)} |X(s)|ds \\ &+ \frac{|\tau^{\alpha-1} - t^{\alpha-1}|}{(\alpha-1)\Gamma(\alpha)} \int_0^1 \left|(1-s)^{\alpha} - \alpha(1-s)^{\alpha-1}\right| |X(s)|ds. \end{aligned}$$

By using the inequality $|a^p - b^p| \le |a-b|^p$, for all $a, b \ge 0$ and $p \in]0,1]$, we deduce that

$$\begin{aligned} &||u_f(\tau) - u_f(t)|| \\ &\le \frac{|\tau - t|^{\alpha-1}}{\Gamma(\alpha)} [\int_0^{\tau} |X(s)|ds + \int_{\tau}^{t} |X(s)|ds + \frac{1+\alpha}{\alpha-1} \int_0^1 |X(s)|ds] \\ &\le [\frac{2\alpha}{(\alpha-1)\Gamma(\alpha)} \int_0^1 |X(s)|ds] |\tau - t|^{\alpha-1}. \end{aligned}$$

The above estimate shows that $\{u_f : f \in S^1_X\}$ is equicontinuous in $\mathcal{C}_E([0,1])$. Since the convexity is obvious and $||u_f(t)|| \leq M_G \int_0^1 |X(s)|ds$ for all $f \in S^1_X$ and for all $t \in [0,1]$, it remains to check the weak compactness property.

We note first that $\mathcal{X}(t) = \int_0^1 G(t,s)X(s)ds$ is convex weakly compact, because $\mathcal{X}(t)$ is the Aumann integral of the convex weakly compact integrably bounded mapping $s \mapsto G(t,s)X(s)$, e.g. [17, 18], so that $\mathcal{X}$ is relatively weakly compact in $\mathcal{C}_E([0,1])$. Let (u_{f_n}) be a sequence of $W^{\alpha,1}_{B,E}([0,1])$-solutions to (16). From Lemma 2.2 we have, for each $n \in \mathbf{N}$,

$$u_{f_n}(t) = \int_0^1 G(t,s)f_n(s)ds, \quad t \in [0,1]$$

with

(i) $I^\beta u_{f_n}(t)|_{t=0} = 0, \ u_{f_n}(1) = \int_0^1 u_{f_n}(t)dt,$
(ii) $u_{f_n} \in W^{\alpha,1}_{B,E}([0,1]),$
(iii) $w\text{-}D^\alpha u_{f_n}(t) = f_n(t)$ a.e.,
(iv) $w\text{-}D^{\alpha-1} u_{f_n}(t) = \int_0^t f_n(s)ds + \frac{1}{\alpha-1}\int_0^1((1-s)^\alpha - \alpha(1-s)^{\alpha-1})f_n(s)ds, \quad t \in [0,1].$

As S^1_X is $\sigma(L^1_E, L^\infty_{E^*})$-compact, by Eberlein-Smulian theorem, we may assume that (f_n) $\sigma(L^1_E, L^\infty_{E^*})$-converges to $f_\infty \in S^1_X$ so that, u_{f_n} weakly converges to u_{f_∞} in $\mathcal{C}_E([0,1])$ where $u_{f_\infty}(t) = \int_0^1 G(t,s)f_\infty(s)ds$, $t \in [0,1]$ with

(i) $I^\beta u_{f_\infty}(t)|_{t=0} = 0, \ u_{f_\infty}(1) = \int_0^1 u_{f_\infty}(t)dt,$
(ii) $u_{f_\infty} \in W^{\alpha,1}_{B,E}([0,1]),$
(iii) $w\text{-}D^\alpha u_{f_\infty}(t) = f_\infty(t)$ a.e.,
(iv) $w\text{-}D^{\alpha-1} u_{f_\infty}(t) = \int_0^t f_\infty(s)ds + \frac{1}{\alpha-1}\int_0^1((1-s)^\alpha - \alpha(1-s)^{\alpha-1})f_\infty(s)ds \quad t \in [0,1].$

Similarly, we show that, for each $t \in [0,1]$

$$\begin{aligned} \text{weak - }\lim_{n\to\infty} w\text{-}D^{\alpha-1} u_{f_n}(t) &= \text{weak - }\lim_{n\to\infty} \int_0^t f_n(s)ds \\ &+ \text{weak - }\lim_{n\to\infty} \frac{1}{\alpha-1}\int_0^1 ((1-s)^\alpha - \alpha(1-s)^{\alpha-1})f_n(s)ds \\ = \int_0^t f_\infty(s)ds + \frac{1}{\alpha-1}\int_0^1 &((1-s)^\alpha - \alpha(1-s)^{\alpha-1})f_\infty(s)ds = D^{\alpha-1}u_{f_\infty}(t). \end{aligned}$$

Thus $w\text{-}D^{\alpha-1}u_{f_n}$ weakly converges to $w\text{-}D^{\alpha-1}u_{f_\infty}$ in $\mathcal{C}_E([0,1])$. ■

Remark. In the course of the proof of Theorem 3.1, we have proven the continuous dependence $S^1_X \to \mathcal{C}_E([0,1])$ of the mappings $f \mapsto u_f$ and $f \mapsto w\text{-}D^{\alpha-1}u_f$ on the convex $\sigma(L^1_E, L^\infty_{E^*})$-compact set S^1_X. This fact has some importance in further applications.

Theorem 3.2. *Let $X : [0,1] \hookrightarrow E$ be a convex weakly compact valued measurable and integrably bounded mapping. Let $F : [0,1] \times E_\sigma \times E_\sigma \hookrightarrow E$ be a convex weakly compact valued mapping satisfying*

(1) F is scalarly $\mathcal{L}_\lambda([0,1]) \otimes \mathcal{B}(E_\sigma) \otimes \mathcal{B}(E_\sigma)$-measurable,[1] *i.e., for each $x^* \in E^*$, the scalar function $\delta^*(x^*, F(.,.,.))$ is $\mathcal{L}_\lambda([0,1]) \otimes \mathcal{B}(E_\sigma) \otimes \mathcal{B}(E_\sigma)$-measurable,*
(2) The mapping $F(t,.,.)$ is scalarly sequentially upper semicontinuous, i.e., for each $t \in [0,1]$ and for each $x^* \in E^*$, the scalar function $\delta^*(x^*, F(t,.,.))$ is sequentially upper semicontinuous on $E_\sigma \times E_\sigma$,
(3) $F(t,x,y) \subset X(t)$ for all $(t,x,y) \in [0,1] \times E \times E$.

Then the $W^{\alpha,1}_{B,E}([0,1])$-solutions set to the (FDI)

$$\begin{cases} w\text{-}D^\alpha u(t) \in F(t, u(t), w\text{-}D^{\alpha-1}u(t)),\ t \in [0,1] \\ I^\beta u(t)|_{t=0} = 0,\ u(1) = \int_0^1 u(t)dt \end{cases} \tag{17}$$

is weakly compact in the space $C_E([0,1])$.

Proof.

Step 1: Existence : For each $h \in S^1_X$ let us define

$$\Phi(h) = \{f \in L^1_E[0,1] : f(t) \in F(t, u_h(t), w\text{-}D^{\alpha-1}u_h(t)),\ a.e.\, t \in [0,1]\} \tag{18}$$

where

$$u_h(t) = \int_0^1 G(t,s)h(s)ds,\ \ t \in [0,1]. \tag{19}$$

Taking account of (1), (2), (3), it is clear that $\Phi(h)$ is nonempty with $\Phi(h) \subset S^1_X$. In fact $\Phi(h)$ is the set of integrable selections of the convex weakly compact valued mapping $K_h(t) := F(t, u_h(t), w\text{-}D^{\alpha-1}u_h(t))$, $t \in [0,1]$. From Lemma 3.3 and (18) it is clear that any fixed point of Φ constitutes a $W^{\alpha,1}_{B,E}([0,1])$-solution to (17). We show that $\Phi : S^1_X \hookrightarrow S^1_X$ is a convex $\sigma(L^1_E, L^\infty_{E^*})$-compact valued upper semicontinuous mapping. By weak compactness, it is enough to show that the graph of Φ is sequentially $\sigma(L^1_E, L^\infty_{E^*})$-closed. Let $f_n \in \Phi(h_n)$ such that (h_n) $\sigma(L^1_E, L^\infty_{E^*})$-converges to $h \in S^1_X$ and (f_n) $\sigma(L^1_E, L^\infty_{E^*})$-converges to $f \in S^1_X$. We need to show that $f \in \Phi(h)$. By (18) and the remark of the proof of Theorem 3.1 we have that $u_{h_n}(t) \to u_h(t)$ and that $w\text{-}D^{\alpha-1}u_{h_n}(t) \to w\text{-}D^{\alpha-1}u_h(t)$ for each $t \in [0,1]$, for E endowed with the weak topology. Now from

$$f_n(t) \in F(t, u_{h_n}(t), D^{\alpha-1}u_{h_n}(t)),\ a.e.\, t \in [0,1]$$

[1] Since E is a separable Banach space, $\mathcal{B}(E_\sigma) = \mathcal{B}(E)$.

we deduce

$$\langle 1_A(t)x^*, f_n(t)\rangle \leq \delta^*(1_A(t)x^*, F(t, u_{h_n}(t), w\text{-}D^{\alpha-1}u_{h_n}(t)))\ a.e.\ t \in [0,1]$$

for each $A \in \mathcal{L}_\lambda([0,1])$ and for each $x^* \in E^*$ so that by integrating

$$\int_A \langle x^*, f_n(t)\rangle dt \leq \int_A \delta^*(x^*, F(t, u_{h_n}(t), w\text{-}D^{\alpha-1}u_{h_n}(t)))dt.$$

Thus we have

$$\begin{aligned}\lim_n \int_A \langle x^*, f_n(t)\rangle dt &= \int_A \langle x^*, f(t)\rangle dt \\ &\leq \limsup_n \int_A \delta^*(x^*, F(t, u_{h_n}(t), w\text{-}D^{\alpha-1}u_{h_n}(t)))dt \\ &\leq \int_A \limsup_n \delta^*(x^*, F(t, u_{h_n}(t), w\text{-}D^{\alpha-1}u_{h_n}(t)))dt \\ &\leq \int_A \delta^*(x^*, F(t, u_h(t), w\text{-}D^{\alpha-1}u_h(t)))dt.\end{aligned}$$

Whence we get

$$\int_A \langle x^*, f(t)\rangle dt \leq \int_A \delta^*(x^*, F(t, u_h(t), w\text{-}D^{\alpha-1}u_h(t)))dt$$

for every $A \in \mathcal{L}_\lambda([0,1])$. Consequently

$$\langle x^*, f(t)\rangle \leq \delta^*(x^*, F(t, u_h(t), w\text{-}D^{\alpha-1}u_h(t)))\ a.e.$$

By taking a dense sequence (e_k^*) in E^* with respect to the Mackey topology $\tau(E^*, E)$, we get

$$\langle e_k^*, f(t)\rangle \leq \delta^*(e_k^*, F(t, u_h(t), w\text{-}D^{\alpha-1}u_h(t)))\ a.e.$$

for all $k \in \mathbf{N}$, by ([17], Prop. III.35) we get finally

$$f(t) \in F(t, u_h(t), w\text{-}D^{\alpha-1}u_h(t))\ a.e.$$

Apply Kakutani-Ky Fan fixed point theorem to the convex $\sigma(L_E^1, L_{E^*}^\infty)$-compact valued upper semicontinuous mapping Φ shows that Φ admits a fixed point, $h \in \Phi(h)$, thus proving the existence of the $W_{B,E}^{\alpha,1}$-solutions to the (FDI) (17).

Step 2: Weak compactness: follows easily from the above arguments and the weak compactness in $\mathcal{C}_E([0,1])$ of

$$\mathcal{X} = \{u_f : [0,1] \to E,\ u_f(t) = \int_0^1 G(t,s)f(s)ds,\ f \in S^1_X,\ t \in [0,1]\}$$

given in Theorem 3.1. ■

In this regard, several variants of the above results can be found in [21] dealing with compact properties of the $W^{\alpha,1}_{B,E}([0,1])$-solutions to fractional differential inclusion (FDI) of the form

$$\begin{cases} w\text{-}D^{\alpha}u(t) \in F(t, u(t), w\text{-}D^{\alpha-1}u(t)),\ \ t \in [0,1] \\ I^{\beta}u(t)|_{t=0} = 0,\ \ u(1) = \int_0^1 u(t)dt \end{cases}$$

where $\alpha \in]1,2], \beta \in [0,\infty[$ are given constants, and $w\text{-}D^{\alpha}$ and $w\text{-}D^{\alpha-1}$ are the fractional w-R.L derivatives of α and $\alpha - 1$ respectively, F is a convex compact valued upper semi continuous mapping.

4 Fractional Differential Inclusion in Pettis Setting

In this section we proceed to the study of fractional differential inclusion (FDI) in the Pettis context of the form

$$\begin{cases} w\text{-}D^{\alpha}u(t) \in F(t, u(t), w\text{-}D^{\alpha-1}u(t)),\ \ t \in [0,1] \\ I^{\beta}u(t)|_{t=0} = 0,\ \ u(1) = \int_0^1 u(t)dt \end{cases} \tag{20}$$

where $\alpha \in]1,2]$, $\beta \in]0,\infty[$ are given constants, and $w\text{-}D^{\alpha}$ and $w\text{-}D^{\alpha-1}$ are the fractional w-R.L derivatives of α and $\alpha - 1$ respectively, F is a convex weakly compact valued mapping.

We need at first some Pettis analogs of the results developed in the preceding section.

Lemma 4.1. *Let $f \in P^1_E([0,1])$ be a Pettis-integrable function. Then the function*

$$u_f(t) = \int_0^1 G(t,s)f(s)ds, \quad t \in [0,1]$$

is the unique $W^{\alpha,1}_{P,E}([0,1])$*-solution to the fractional differential equation (FDE)*

$$\begin{cases} w\text{-}D^{\alpha}u(t) = f(t), \ t \in [0,1] \\ I^{\beta}u(t)|_{t=0} = 0, \ u(1) = \int_0^1 u(t)dt \end{cases}$$

We begin with a crucial lemma.

Lemma 4.2. *Let* $X : [0,1] \hookrightarrow E$ *be a convex weakly compact valued Pettis-integrable multifunction. Then the* $W^{\alpha,1}_{P,E}([0,1])$*-solutions set* $\mathcal{X}$ *to the (FDI)*

$$\begin{cases} w\text{-}D^{\alpha}u(t) \in X(t), \ t \in [0,1] \\ I^{\beta}u(t)|_{t=0} = 0, \ u(1) = \int_0^1 u(t)dt \end{cases} \tag{21}$$

is bounded, convex, equicontinuous, weakly compact in the Banach space $\mathcal{C}_E([0,1])$.

Proof.

Step 1. Let us recall that the set S_X^{Pe} of all Pettis-integrable selections of X is nonempty and sequentially compact for the topology of pointwise convergence on $L^{\infty} \otimes E^*$ and the multivalued integral

$$\int_0^1 X(t)\,dt = \{\int_0^1 f(t)\,dt : f \in S_X^{Pe}\}$$

is convex and weakly compact in E (see [13, 17]).

Step 2. In view of Lemma 4.1, the $W^{\alpha,1}_{P,E}([0,1])$-solutions set $\mathcal{X}$ to (21) is characterized by

$$\mathcal{X} = \{u_f : [0,1] \to E \mid u_f(t) = \int_0^1 G(t,s)f(s)\,ds, \ \forall t \in [0,1]; \ f \in S_X^{Pe}\}.$$

Recall that X is Pettis-integrable and the set $\{\delta^*(x^*, X(.)) : x^* \in \overline{B}_{E^*}\}$ is uniformly integrable in $L^1_{\mathbf{R}}([0,1])$. Now it is not difficult to show that $\mathcal{X}$ is equicontinuous in $\mathcal{C}_E([0,1])$. Indeed, let $x^* \in \overline{B}_{E*}$ and let $\tau, t \in [0,1]$, we have the estimate

$$|\langle x^*, u_f(\tau) - u_f(t)\rangle| \leq \int_0^1 |G(t,s) - G(\tau,s)|\,|\delta^*(x^*, X(s))|ds$$

$$\leq \frac{1}{\Gamma(\alpha)} \int_0^{\tau} \left|(\tau - s)^{\alpha-1} - (t-s)^{\alpha-1}\right| |\delta^*(x^*, X(s))|ds$$

$$+ \int_{\tau}^{t} \frac{(t-s)^{\alpha-1}}{\Gamma(\alpha)} |\delta^*(x^*, X(s))|ds$$

$$+ \frac{|\tau^{\alpha-1} - t^{\alpha-1}|}{(\alpha-1)\Gamma(\alpha)} \int_0^1 \left|(1-s)^{\alpha} - \alpha(1-s)^{\alpha-1}\right| |\delta^*(x^*, X(s))|ds.$$

By using the inequality $|a^p - b^p| \leq |a-b|^p$, for all $a, b \geq 0$ and $p \in]0,1]$, we deduce that

$$
\begin{aligned}
&|\langle x^*, u_f(\tau) - u_f(t)\rangle| \\
&\leq \frac{|t-\tau|^{\alpha-1}}{\Gamma(\alpha)}[\sup_{x^*\in\overline{B}_{E^*}} \int_0^\tau |\delta^*(x^*, X(s))|ds + \sup_{x^*\in\overline{B}_{E^*}} \int_\tau^t |\delta^*(x^*, X(s))|ds \\
&\quad + \sup_{x^*\in\overline{B}_{E^*}} \frac{1+\alpha}{\alpha-1}\int_0^1 |\delta^*(x^*, X(s))|ds] \\
&\leq [\frac{2\alpha}{(\alpha-1)\Gamma(\alpha)} \sup_{x^*\in\overline{B}_{E^*}} \int_0^1 |\delta^*(x^*, X(s))|ds]|t-\tau|^{\alpha-1}.
\end{aligned}
$$

The above estimate shows that $\mathcal{X}$ is equicontinuous in $\mathcal{C}_E([0,1])$. Further, for each $t \in [0,1]$, the set $\mathcal{X}(t)$ is relatively weakly compact in E because it is included in the weakly compact set $\int_0^1 G(t,s)X(s)\,ds$. We claim that $\mathcal{X}$ is weakly compact in $\mathcal{C}_E([0,1])$. Let (f_n) be a sequence in S_Γ^{Pe}. We note that S_X^{Pe} is sequentially compact for the topology of pointwise convergence on $L^\infty \otimes E^*$. We note that the sequence (u_{f_n}) is sequentially weakly compact in $\mathcal{C}_E([0,1])$. Indeed we can extract from (f_n) a sequence (f_m) such that (f_m) $\sigma(P_E^1, L^\infty \otimes E^*)$-converges to a function $f \in S_X^{Pe}$. So, for every $x^* \in E^*$ and for every $t \in [0,1]$, we have

$$
\begin{aligned}
&\lim_{m\to+\infty} \int_0^1 \langle G(t,s)x^*, f_m(s)\rangle\, ds \\
&= \lim_{m\to+\infty} \langle x^*, \int_0^1 G(t,s)f_m(s)\rangle\, ds \\
&= \int_0^1 \langle G(t,s)x^*, f(s)\rangle\, ds = \langle x^*, \int_0^1 G(t,s)f(s)\, ds\rangle.
\end{aligned}
$$

This means that u_{f_m} converges weakly in $\mathcal{C}_E([0,1])$ to u_f. This shows the weak compactness of $\mathcal{X}$ in $\mathcal{C}_E([0,1])$. At this point, it is worth to mention that the sequence $(w\text{-}D^{\alpha-1}u_{f_m})$ weakly converges to $w\text{-}D^{\alpha-1}u_f$ in $\mathcal{C}_E([0,1])$, by using the $\sigma(P_E^1, L^\infty \otimes E^*)$-convergence of (f_m) and Lemma 2.5(iv).

Remark.

(1) In the course of the proof of Lemma 4.2, we have proven the continuous dependence of the mappings $f \mapsto u_f$ and $f \mapsto w\text{-}D^{\alpha-1}u_f$ on the convex $\sigma(P_E^1, L^\infty \otimes E^*)$-compact set S_X^{Pe}. This fact has some importance in further applications.

(2) It is worth to mention that $\{w\text{-}D^{\alpha-1}u_f : f \in S_X^{Pe}\}$ is also bounded, convex, equicontinuous, weakly compact in $\mathcal{C}_E([0, 1])$ and the multivalued integral

$$\int_0^t X(s)ds + \frac{1}{\alpha-1}\int_0^1 ((1-s)^\alpha - \alpha(1-s)^{\alpha-1})X(s)ds.$$

is weakly compact in E.

Now we proceed to the existence of solutions in $W_{P,E}^{\alpha,1}([0, 1])$ to the(FDI)

$$\begin{cases} w\text{-}D^\alpha u(t) \in F(t, u(t), w\text{-}D^{\alpha-1}u(t)), \ t \in [0, 1] \\ I^\beta(t)|_{t=0} = 0, \ u(1) = \int_0^1 u(t)dt \end{cases}$$

Theorem 4.1. *Let $X : [0, 1] \hookrightarrow E$ be a convex weakly compact valued Pettis-integrable multifunction. Let $F : [0, 1] \times E_\sigma \times E_\sigma \hookrightarrow E$ be a convex weakly compact valued mapping satisfying*

(1) F is scalarly $\mathcal{L}_\lambda([0, 1]) \otimes \mathcal{B}(E_\sigma) \otimes \mathcal{B}(E_\sigma)$-measurable, i.e., for each $x^* \in E^*$, the scalar function $\delta^*(x^*, F(., ., .))$ is $\mathcal{L}_\lambda([0, 1]) \otimes \mathcal{B}(E_\sigma) \otimes \mathcal{B}(E_\sigma)$-measurable,

(2) The mapping $F(t, ., .)$ is scalarly sequentially upper semicontinuous, i.e., for each $t \in [0, 1]$ and for each $x^* \in E^*$, the scalar function $\delta^*(x^*, F(t, ., .))$ is sequentially upper semicontinuous on $E_\sigma \times E_\sigma$,

(3) $F(t, x, y) \subset X(t)$ for all $(t, x, y) \in [0, 1] \times E \times E$.

Then the $W_{P,E}^{\alpha,1}([0, 1])$-solutions set to the (FDI)

$$\begin{cases} w\text{-}D^\alpha u(t) \in F(t, u(t), w\text{-}D^{\alpha-1}u(t)), \ t \in [0, 1] \\ I^\beta u(t)|_{t=0} = 0, \ u(1) = \int_0^1 u(t)dt \end{cases} \tag{22}$$

is weakly compact in the space $\mathcal{C}_E([0, 1])$.

Proof. Taking the above stated results into account, a mapping $u : [0, 1] \to E$ is a $W_{P,E}^{\alpha,1}([0, 1])$-solution of the above (FDI) iff there exists $f \in S_X^{Pe}$ such that

$$u(t) = u_f(t) = \int_0^1 G(t, s)f(s)ds, \ \forall t \in [0, 1]$$

and such that

$$D^\alpha u_f(t) = f(t) \in F(t, u_f(t), D^{\alpha-1}u_f(t)) \ \forall t \in [0, 1].$$

Let $\mathcal{X}$ be defined in Lemma 4.2. For each $u \in \mathcal{X}$, let us set

$$\Phi(u) = \{v \in \mathcal{X} : w\text{-}D^\alpha v(t) \in F(t, u(t), w\text{-}D^{\alpha-1}u(t) \ a.e.\}.$$

It is clear that $\Phi(u)$ is nonempty, convex weakly compact in $\mathcal{X}$. We need to check that the graph of Φ is sequentially weakly closed in $\mathcal{X} \times \mathcal{X}$. Let $v_n \in \Phi(u_n)$ such that $u_n \to u$ and $v_n \to v$ in $\mathcal{X}$. By the sequential $\sigma(P_E^1, L^\infty \otimes E^*)$-compactness of S_X^{Pe} and Lemma 4.2, we may assume that $u_n(t) = u_{f_n}(t) \to u(t) = u_f(t)$ in E_σ for each $t \in [0,1]$, $w\text{-}D^{\alpha-1}u_n(t) = w\text{-}D^{\alpha-1}u_{f_n}(t) \to w\text{-}D^{\alpha-1}u(t)$ in E_σ for each $t \in [0,1]$, and $w\text{-}D^\alpha u_n = f_n$ $\sigma(P_E^1, L^\infty \otimes E^*)$-converges to $w\text{-}D^\alpha u = f$ and similarly $v_n(t) = u_{g_n}(t) \to v(t) = v_g(t)$ in E_σ for each $t \in [0,1]$, $w\text{-}D^{\alpha-1}v_n(t) = w\text{-}D^{\alpha-1}u_{g_n}(t) \to w\text{-}D^{\alpha-1}v(t)$ in E_σ for each $t \in [0,1]$, and $w\text{-}D^\alpha v_n = g_n$ $\sigma(P_E^1, L^\infty \otimes E^*)$-converges to $w\text{-}D^\alpha v = g$ with

$$g_n(t) \in F(t, u_{f_n}(t), w\text{-}D^{\alpha-1}u_{f_n}(t)) \ a.e.$$

From the above inclusion, we have

$$\langle h \otimes x^*, g_n(t)\rangle \le \delta^*(h \otimes x^*, F(t, u_{f_n}(t), w\text{-}D^{\alpha-1}u_{f_n}(t))) \ a.e.$$

where $h \in L^\infty([0,1,dt)$ and $x^* \in E^*$. By integrating we get

$$\int_0^1 \langle h \otimes x^*, g_n(t)\rangle dt \le \int_0^1 \delta^*(h \otimes x^*, F(t, u_{f_n}(t), w\text{-}D^{\alpha-1}u_{f_n}(t)))dt.$$

Whence we deduce that

$$\begin{aligned}
\lim_{n\to\infty} \int_0^1 \langle h \otimes x^*, g_n(t)\rangle dt &= \int_0^1 \langle h \otimes x^*, g(t)\rangle dt \\
\le \limsup_n \int_0^1 &\delta^*(h \otimes x^*, F(t, u_{f_n}(t), w\text{-}D^{\alpha-1}u_{f_n}(t)))dt \\
\le \int_0^1 &\limsup_n \delta^*(h \otimes x^*, F(t, u_{f_n}(t), w\text{-}D^{\alpha-1}u_{f_n}(t)))dt \\
&\le \int_0^1 \delta^*(h \otimes x^*, F(t, u_f(t), w\text{-}D^{\alpha-1}u_f(t)))dt
\end{aligned}$$

so that, in particular, by taking $h = 1_A$ with $A \in \mathcal{L}_\lambda([0,1]$ we get

$$\int_A \langle x^*, g(t)\rangle dt \le \int_A \delta^*(x^*, F(t, u_f(t), w\text{-}D^{\alpha-1}u_f(t)))dt$$

so that

$$\langle x^*, g(t)\rangle \le \delta^*(x^*, F(t, u_f(t), w\text{-}D^{\alpha-1}u_f(t))) \ a.e.$$

By taking a dense sequence (e_k^*) with respect to the Mackey topology, we get

$$\langle e_k^*, g(t)\rangle \leq \delta^*(e_k^*, F(t, u_f(t), w\text{-}D^{\alpha-1}u_f(t))) \ a.e.$$

By virtue of ([17], Prop. III.35) we conclude that

$$g(t) \in F(t, u_f(t), w\text{-}D^{\alpha-1}u_f(t)) \ a.e.$$

thus proving that the graph of Φ is weakly compact in $\mathcal{X} \times \mathcal{X}$. ■

We finish this section by providing some new variants of the above stated theorems.

The following lemma is useful for our purpose.

Lemma 4.3. *Let* $\alpha \in]1, 2]$, $b \in E$ *and let* $f \in L_E^1([0, 1])$. *Then the mapping* $u_f : [0, 1] \to E$ *defined by*

$$u_f(t) = \frac{1}{\Gamma(\alpha)} \int_0^t (t-s)^{\alpha-1} f(s)ds + \frac{b}{\Gamma(\alpha)} t^{\alpha-1}, t \in [0, 1]$$

is the unique $W_{B,E}^{\alpha,1}([0, 1])$*-solution to the boundary problem*

$$w\text{-}D^\alpha u(t) = f(t), \ t \in [0, 1],$$
$$u(0) = 0, \ w\text{-}D^{\alpha-1}u(0) = b.$$

Proof. It is worth to note first that u_f is continuous by Lemma 2.1 with $u_f(0) = 0$. We note that

$$w\text{-}D^\alpha u_f(t) = w\text{-}D^\alpha I^\alpha f(t) + \frac{b}{\Gamma(\alpha)} D^\alpha t^{\alpha-1}$$

From the classical calculus of fractional order integral and R. L derivatives, we have that $D^\alpha t^{\alpha-1} = 0$ and $w\text{-}D^\alpha I^\alpha f(t) = f(t)$ so that $w\text{-}D^\alpha u_f(t) = f(t)$. Similarly we have $w\text{-}D^{\alpha-1}(I^\alpha f(t)) = I^1 f(t) = \int_0^t f(s)ds$ and $D^{\alpha-1}t^{\alpha-1} = \Gamma(\alpha)$ so that $w\text{-}D^{\alpha-1}u_f(t) = \int_0^t f(s)ds + b$ and that $w\text{-}D^{\alpha-1}u_f(0) = b$. ■

Lemma 4.4. *Let* $X : [0, 1] \hookrightarrow E$ *be a convex compact valued measurable and integrably bounded multifunction. Then the* $W_{B,E}^{\alpha,1}([0, 1])$*-solutions set to the (FDI)*

$$\begin{cases} w\text{-}D^\alpha u(t)) \in X(t), \ t \in [0, 1] \\ u(0) = 0, \ w\text{-}D^{\alpha-1}u(0) = b. \end{cases} \tag{23}$$

is compact in the space $C_E([0, 1])$ *endowed with the topology of uniform convergence.*

Proof. By virtue of Lemma 4.3, the $W^{\alpha,1}_{B,E}([0,1])$-solution set $\mathcal{X}$ of (23) is characterized by

$$\mathcal{X} = \{u_f : [0,1] \to E,\ u_f(t) = \frac{1}{\Gamma(\alpha)} \int_0^t (t-s)^{\alpha-1} f(s)ds + \frac{b}{\Gamma(\alpha)} t^{\alpha-1},\quad f \in S^1_X,\ t \in [0,1]\}$$

Claim. $\mathcal{X}$ is compact in $C_E([0,1])$.

Let (u_{f_n}) be a sequence of $W^{\alpha,1}_{B,E}([0,1])$-solutions to (23). As S^1_X is $\sigma(L^1_E, L^\infty_{E*})$-compact, by Eberlein-Smulian theorem, we may assume that (f_n) $\sigma(L^1_E, L^\infty_{E*})$-converges to $f_\infty \in S^1_X$. From Lemma 4.3 we have, for each $n \in \mathbf{N}$,

$$u_{f_n}(t) = \frac{1}{\Gamma(\alpha)} \int_0^t (t-s)^{\alpha-1} f_n(s)ds + \frac{b}{\Gamma(\alpha)} t^{\alpha-1}$$

with

(i) $u_{f_n}(0) = 0, w\text{-}D^{\alpha-1}u_{f_n}(0) = b$,
(ii) $u_{f_n} \in W^{\alpha,1}_{B,E}([0,1])$,
(iii) $w\text{-}D^{\alpha}u_{f_n}(t) = f_n(t)$ a.e.,
(iv) $w\text{-}D^{\alpha-1}u_{f_n}(t) = b + \int_0^t f_n(s)ds$.

On the other hand, it is not difficult to show that $\{u_{f_n} : n \in \mathbf{N}\}$ is equicontinuous in $C_E([0,1])$. Further, for each $t \in [0,1]$ $\{u_{f_n}(t) : n \in \mathbf{N}\}$ is relatively compact because it is included in the norm compact set

$$\frac{1}{\Gamma(\alpha)} \int_0^t (t-s)^{\alpha-1} X(s)ds + \frac{b}{\Gamma(\alpha)} t^{\alpha-1}$$

(see e.g. [11, 17, 18]). So by Ascoli's theorem, $\{u_{f_n} : n \in \mathbf{N}\}$ is relatively compact in $C_E([0,1])$ and we may assume that (u_{f_n}) converges uniformly to a function $u_\infty \in C_E([0,1])$. We have, for each $t \in [0,1]$,

$$\begin{aligned} u_\infty(t) &= \lim_{n\to\infty} u_{f_n}(t) \\ &= \lim_{n\to\infty} \frac{1}{\Gamma(\alpha)} \int_0^t (t-s)^{\alpha-1} f_n(s)ds + \frac{b}{\Gamma(\alpha)} t^{\alpha-1} \\ &= \frac{1}{\Gamma(\alpha)} \int_0^t (t-s)^{\alpha-1} f_\infty(s)ds + \frac{bt^{\alpha-1}}{\Gamma(\alpha)} t^{\alpha-1} = u_{f_\infty}(t). \end{aligned}$$

Thus $u_\infty = u_{f_\infty}$ so that by using Lemma 4.3

(i) $u_{f_\infty}(0) = 0, w\text{-}D^{\alpha-1}u_{f_\infty}(0) = b$
(ii) $u_{f_\infty} \in W^{\alpha,1}_{B,E}([0,1])$,

(iii) $w\text{-}D^{\alpha}u_{f_\infty}(t) = f_\infty(t)$ a.e.,
(iv) $w\text{-}D^{\alpha-1}u_{f_\infty}(t) = b + \int_0^t f_\infty(s)ds$.

Similarly, we show that, for each $t \in [0, 1]$

$$\lim_{n\to\infty} w\text{-}D^{\alpha-1}u_{f_n}(t) = \lim_{n\to\infty} b + \int_0^t f_n(s)ds = b + \int_0^t f_\infty(s)ds = w\text{-}D^{\alpha-1}u_{f_\infty}(t).$$

■

Remark. In the course of the proof of Lemma 4.4, we have proven the continuous dependence of the mappings $f \mapsto u_f$ and $f \mapsto w\text{-}D^{\alpha-1}u_f$ on the convex $\sigma(L^1_E, L^\infty_{E^*})$-compact set S^1_X. This fact has some importance in further applications.

Theorem 4.2. *Let $X : [0,1] \hookrightarrow E$ be a convex compact valued measurable and integrably bounded mapping. Let $F : [0,1] \times E \times E \hookrightarrow E$ be a convex compact valued mapping satisfying*

(1) F is scalarly $\mathcal{L}_\lambda([0,1]) \otimes \mathcal{B}(E) \otimes \mathcal{B}(E)$-measurable, i.e., for each $x^* \in E^*$, the scalar function $\delta^*(x^*, F(.,.,.))$ is $\mathcal{L}_\lambda([0,1]) \otimes \mathcal{B}(E) \otimes \mathcal{B}(E)$-measurable,
(2) The mapping $F(t,.,.)$ is scalarly upper semicontinuous, i.e., for each $t \in [0,1]$ and for each $x^* \in E^*$, the scalar function $\delta^*(x^*, F(t,.,.))$ is upper semicontinuous on $E \times E$,
(3) $F(t,x,y) \subset X(t)$ for all $(t,x,y) \in [0,1] \times E \times E$.

Then the $W^{\alpha,1}_{B,E}([0,1])$-solutions set to the (FDI)

$$\begin{cases} w\text{-}D^{\alpha}u(t) \in F(t, u(t), w\text{-}D^{\alpha-1}u(t)),\ t \in [0,1] \\ u(0) = 0,\ w\text{-}D^{\alpha-1}u(0) = b \end{cases} \tag{24}$$

is compact in the space $C_E([0,1])$ endowed with the topology of uniform convergence.

Proof.

Step 1: Existence: For each $h \in S^1_X$, let us define

$$\Phi(h) = \{f \in L^1_E[0,1] : f(t) \in F(t, u_h(t), w\text{-}D^{\alpha-1}u_h(t)),\ a.e.\ t \in [0,1]\} \tag{25}$$

where

$$u_h(t) = \frac{1}{\Gamma(\alpha)} \int_0^t (t-s)^{\alpha-1}h(s)ds + \frac{b}{\Gamma(\alpha)}t^{\alpha-1}. \tag{26}$$

Taking account of (1), (2), (3), it is clear that $\Phi(h)$ is nonempty with $\Phi(h) \subset S^1_X$. In fact $\Phi(h)$ is the set of integrable selections of the convex compact valued mapping $K_h(t) := F(t, u_h(t), w\text{-}D^{\alpha-1}u_h(t))$, $t \in [0,1]$. From Lemma 4.3 and (25) it is clear that any fixed point of Φ constitutes a $W^{\alpha,1}_{B,E}([0,1])$-solution to (24).

We show that $\Phi : S^1_X \hookrightarrow S^1_X$ is a convex $\sigma(L^1_E, L^\infty_{E^*})$-compact valued upper semicontinuous mapping. By compactness, it is enough to show that the graph of Φ is sequentially $\sigma(L^1_E, L^\infty_{E^*})$-closed. Let $f_n \in \Phi(h_n)$ such that (h_n) $\sigma(L^1_E, L^\infty_{E^*})$-converges to $h \in S^1_X$ and (f_n) $\sigma(L^1_E, L^\infty_{E^*})$-converges to $f \in S^1_X$. We need to show that $f \in \Phi(h)$. By Lemma 4.4 and its remark, we have that $u_{h_n}(t) \to u_h(t)$ and that $w\text{-}D^{\alpha-1}u_{h_n}(t) \to w\text{-}D^{\alpha-1}u_h(t)$ for each $t \in [0,1]$, for E endowed with the strong topology. Now from

$$f_n(t) \in F(t, u_{h_n}(t), w\text{-}D^{\alpha-1}u_{h_n}(t)), \ a.e.\ t \in [0,1]$$

we deduce

$$\langle 1_A(t)x^*, f_n(t)\rangle \leq \delta^*(1_A(t)x^*, F(t, u_{h_n}(t), w\text{-}D^{\alpha-1}u_{h_n}(t)))\ a.e.\ t \in [0,1]$$

for each $A \in \mathcal{L}_\lambda([0,1])$ and for each $x^* \in E^*$ so that by integrating

$$\int_A \langle x^*, f_n(t)\rangle dt \leq \int_A \delta^*(x^*, F(t, u_{h_n}(t), w\text{-}D^{\alpha-1}u_{h_n}(t)))dt.$$

Thus we have

$$\begin{aligned}
\lim_n \int_A \langle x^*, f_n(t)\rangle dt &= \int_A \langle x^*, f(t)\rangle dt \\
&\leq \limsup_n \int_A \delta^*(x^*, F(t, u_{h_n}(t), w\text{-}D^{\alpha-1}u_{h_n}(t)))dt \\
&\leq \int_A \limsup_n \delta^*(x^*, F(t, u_{h_n}(t), w\text{-}D^{\alpha-1}u_{h_n}(t)))dt \\
&\leq \int_A \delta^*(x^*, F(t, u_h(t), w\text{-}D^{\alpha-1}u_h(t)))dt.
\end{aligned}$$

Whence we get

$$\int_A \langle x^*, f(t)\rangle dt \leq \int_A \delta^*(x^*, F(t, u_h(t), w\text{-}D^{\alpha-1}u_h(t)))dt$$

for every $A \in \mathcal{L}_\lambda([0,1])$. Consequently

$$\langle x^*, f(t)\rangle \leq \delta^*(x^*, F(t, u_h(t), w\text{-}D^{\alpha-1}u_h(t)))\ \ a.e.$$

By taking a dense sequence (e^*_k) in E^* with respect to the Mackey topology $\tau(E^*, E)$, we get

$$\langle e^*_k, f(t)\rangle \leq \delta^*(e^*_k, F(t, u_h(t), w\text{-}D^{\alpha-1}u_h(t)))\ \ a.e.$$

for all $k \in \mathbf{N}$, by ([17], Prop. III.35) we get finally

$$f(t) \in F(t, u_h(t), w\text{-}D^{\alpha-1}u_h(t)) \ a.e.$$

Apply Kakutani-Ky Fan fixed point theorem to the convex $\sigma(L^1_E, L^\infty_{E^*})$-compact valued upper semicontinuous mapping Φ shows that Φ admits a fixed point, $h \in \Phi(h)$, thus proving the existence of the $W^{\alpha,1}_{B,E}([0,1])$-solutions to the (FDI) (24).

Step 2: Compactness: follows easily from the above arguments and the compactness of

$$\mathcal{X} = \{u_f : [0,1] \to E,\ u_f(t) = \frac{1}{\Gamma(\alpha)} \int_0^t (t-s)^{\alpha-1} f(s) ds + \frac{b}{\Gamma(\alpha)} t^{\alpha-1}, \quad f \in S^1_X,\ t \in [0,T]\}$$

■

Now we proceed to further variants of the preceding results.

We begin with a lemma.

Lemma 4.5. *Let $X : [0,1] \hookrightarrow E$ be a convex compact valued Pettis-integrable multifunction. Then the $W^{\alpha,1}_{P,E}([0,1])$-solutions set $\mathcal{X}$ to the (FDI)*

$$\begin{cases} w\text{-}D^\alpha u(t) \in X(t), & t \in [0,1] \\ u(0) = 0, w\text{-}D^{\alpha-1}u(0) = b \end{cases} \tag{27}$$

is compact in the Banach space $\mathcal{C}_E([0,1])$ of all continuous mappings from $[0,1]$ into E endowed with the topology of uniform convergence.

Proof.

Step 1. Let us recall that the set S^{Pe}_X of all Pettis-integrable selections of X is nonempty and sequentially compact for the topology of pointwise convergence on $L^\infty \otimes E^*$ and the multivalued integral

$$\int_0^1 X(t)\, dt = \{\int_0^1 f(t)\, dt : f \in S^{Pe}_X\}$$

is convex and compact in E (see [13, 17]).

Step 2. The $W^{\alpha,1}_{P,E}([0,1])$-solutions set $\mathcal{X}$ to (27) is characterized by

$$\mathcal{X} = \{u_f : [0,1] \to E : u_f(t) = \frac{1}{\Gamma(\alpha)} \int_0^t (t-s)^{\alpha-1} f(s) ds + \frac{b}{\Gamma(\alpha)} t^{\alpha-1}, \quad \forall t \in [0,1], \quad f \in S^{Pe}_X\}$$

Recall that X is Pettis-integrable and the set $\{\delta^*(x^*, X(.)) : x^* \in \overline{B}_{E^*}\}$ is uniformly integrable in $L^1_{\mathbf{R}}([0,1])$. Now we show first that $\mathcal{X}$ is equicontinuous in $\mathcal{C}_E([0,1])$. Indeed, let $x^* \in \overline{B}_{E*}$ and let $t_n < t$ with $t_n \to t$ in $[0,1]$. By using $|a^p - b^p| \le |a-b|^p$, for all $a, b \ge 0$ and $p \in [0,1]$, we have the estimate

$$\sup_{x^* \in \overline{B}_{E*}} |\langle x^*, u_f(t_n) - u_f(t)\rangle| \le \frac{|t_n^{\alpha-1} - t^{\alpha-1}|}{\Gamma(\alpha)}||b||$$

$$+ \sup_{x^* \in \overline{B}_{E*}} |\langle x^*, \int_0^{t_n} \frac{(t_n - s)^{\alpha-1}}{\Gamma(\alpha)} f(s)ds\rangle - \langle x^*, \int_0^{t} \frac{(t - s)^{\alpha-1}}{\Gamma(\alpha)} f(s)ds\rangle|$$

$$+ \sup_{x^* \in \overline{B}_{E*}} |\langle x^*, \int_0^{t_n} \frac{(t_n - s)^{\alpha-1}}{\Gamma(\alpha)} f(s)ds\rangle - \langle x^*, \int_0^{t_n} \frac{(t - s)^{\alpha-1}}{\Gamma(\alpha)} f(s)ds\rangle|$$

$$+ \sup_{x^* \in \overline{B}_{E*}} |\langle x^*, \int_{t_n}^{t} \frac{(t - s)^{\alpha-1}}{\Gamma(\alpha)} f(s)ds\rangle|$$

$$\le \frac{|t_n - t|^{\alpha-1}}{\Gamma(\alpha)}||b|| + \frac{|t_n - t|^{\alpha-1}}{\Gamma(\alpha)} \sup_{x^* \in \overline{B}_{E*}} \int_0^{t_n} |\delta^*(x^*, X(s))|ds$$

$$+ \sup_{x^* \in \overline{B}_{E*}} \int_{t_n}^{t} \frac{(t-s)^{\alpha-1}}{\Gamma(\alpha)} |\delta^*(x^*, X(s))|ds$$

$$\le \frac{|t_n - t|^{\alpha-1}}{\Gamma(\alpha)}||b|| + \frac{|t_n - t|^{\alpha-1}}{\Gamma(\alpha)} \sup_{x^* \in \overline{B}_{E*}} \int_0^{1} |\delta^*(x^*, X(s))|ds$$

so that the second member goes to 0 when $t_n \to t$. The case where $t < t_n$ with $t_n \to t$ in $[0,1]$ can treated similarly. Hence $\mathcal{X}$ is bounded equicontinuous in $\mathcal{C}_E([0,1])$. Further, for each $t \in [0,1]$, the set $\mathcal{X}(t)$ is relatively compact in E because it is included in the compact set

$$\frac{1}{\Gamma(\alpha)} \int_0^t (t-s)^{\alpha-1} X(s)ds + \frac{b}{\Gamma(\alpha)} t^{\alpha-1}.$$

We claim that $\mathcal{X}$ is compact in $\mathcal{C}_E([0,1])$. Let (f_n) be a sequence in S^{Pe}_{Γ}. By Ascoli's theorem, $\{u_{f_n} : n \in \mathbf{N}\}$ is relatively compact in $C_E([0,1])$ and we may assume that (u_{f_n}) converges uniformly to a function $u_\infty \in C_E([0,1])$. We note that S^{Pe}_X is sequentially compact for the topology of pointwise convergence on

$L^\infty \otimes E^*$. So we can extract from the sequence (f_n) a sequence (f_m) such that (f_m) $\sigma(P^1_E, L^\infty \otimes E^*)$-converges to a function $f \in S^{Pe}_X$. We have, for each $t \in [0, 1]$,

$$u_\infty(t) = \lim_{n\to\infty} u_{f_n}(t)$$

$$= \lim_{n\to\infty} \frac{1}{\Gamma(\alpha)} \int_0^t (t-s)^{\alpha-1} f_n(s)ds + \frac{b}{\Gamma(\alpha)} t^{\alpha-1}$$

$$= \frac{1}{\Gamma(\alpha)} \int_0^t (t-s)^{\alpha-1} f_\infty(s)ds + \frac{b}{\Gamma(\alpha)} t^{\alpha-1} = u_{f_\infty}(t).$$

Thus $u_\infty = u_{f_\infty}$. This means that u_{f_m} converges in $\mathcal{C}_E([0,1])$ to u_{f_∞} and shows the compactness of $\mathcal{X}$ in $\mathcal{C}_E([0,1])$. At this point, it is worth to mention that the sequence $(w\text{-}D^{\alpha-1}u_{f_m})(t)$ converges to $w\text{-}D^{\alpha-1}u_{f_\infty}(t)$ in E, for E endowed with the strong topology by using the $\sigma(P^1_E, L^\infty \otimes E^*)$- convergence of (f_m) and the **norm** compactness of $\int_0^t X(s)ds$. ■

Remark.

(1) In the course of the proof of Lemma 4.5, we have proven the continuous dependence of the mappings $f \mapsto u_f$ and $f \mapsto w\text{-}D^{\alpha-1}u_f$ on the convex $\sigma(P^1_E, L^\infty \otimes E^*)$-compact set S^{Pe}_X. This fact has some importance in further applications.
(2) It is worth to mention that $\{w\text{-}D^{\alpha-1}u_f : f \in S^{Pe}_X\}$ is also compact in $\mathcal{C}_E([0,1])$.

Now we proceed to the existence of solutions in $W^{\alpha,1}_{P,E}([0,1])$ for the(FDI)

$$\begin{cases} w\text{-}D^\alpha u(t) \in F(t, u(t), w\text{-}D^{\alpha-1}u(t)), \ \ t \in [0,1] \\ u(0) = 0, \ \ w\text{-}D^{\alpha-1}u(0) = b \end{cases} \tag{28}$$

Theorem 4.3. *Let $X : [0,1] \hookrightarrow E$ be a convex compact valued Pettis-integrable multifunction. Let $F : [0,1] \times E \times E \hookrightarrow E$ be a convex compact valued mapping satisfying*

(1) F is scalarly $\mathcal{L}_\lambda([0,1]) \otimes \mathcal{B}(E) \otimes \mathcal{B}(E)$-measurable, i.e., for each $x^* \in E^*$, the scalar function $\delta^*(x^*, F(.,.,.))$ is $\mathcal{L}_\lambda([0,1]) \otimes \mathcal{B}(E) \otimes \mathcal{B}(E)$-measurable,
(2) The mapping $F(t,.,.)$ is scalarly upper semicontinuous, that is, for each $t \in [0,1]$ and for each $x^* \in E^*$, the scalar function $\delta^*(x^*, F(t,.,.))$ is upper semicontinuous on $E \times E$,
(3) $F(t,x,y) \subset X(t)$ for all $(t,x,y) \in [0,1] \times E \times E$.

Then the $W^{\alpha,1}_{P,E}([0,1])$-solutions set to the (FDI)

$$\begin{cases} w\text{-}D^\alpha u(t) \in F(t, u(t), w\text{-}D^{\alpha-1}u(t)), \ t \in [0,1] \\ u(0) = 0, w\text{-}D^{\alpha-1}u(0) = b \end{cases} \tag{29}$$

is compact in the space $\mathcal{C}_E([0,1])$.

Proof. A mapping $u : [0,1] \to E$ is a $W^{\alpha,1}_{P,E}([0,1])$-solutions to (29) iff there exists $f \in S^{Pe}_X$ such that

$$u(t) = u_f(t) = \frac{1}{\Gamma(\alpha)} \int_0^t (t-s)^{\alpha-1} f(s)ds + \frac{b}{\Gamma(\alpha)} t^{\alpha-1}, \ \ \forall t \in [0,1]$$

and such that

$$w\text{-}D^{\alpha} u_f(t) = f(t) \in F(t, u_f(t), w\text{-}D^{\alpha-1} u_f(t)) \ \ \forall t \in [0,1].$$

Let $\mathcal{X}$ be given in Lemma 4.4. For each $u \in \mathcal{X}$, let us set

$$\Phi(u) = \{v \in \mathcal{X} : w\text{-}D^{\alpha} v(t) \in F(t, u(t), w\text{-}D^{\alpha-1} u(t) \ a.e.\}.$$

It is clear that $\Phi(u)$ is nonempty, convex compact in $C_E([0,1])$. We need to check that the graph of Φ is closed in $\mathcal{X} \times \mathcal{X}$. Let $v_n \in \Phi(u_n)$ such that $u_n \to u$ and $v_n \to v$ in $\mathcal{X}$. By the sequential $\sigma(P^1_E, L^\infty \otimes E^*)$-compactness of S^{Pe}_X, we may assert that $u_n(t) = u_{f_n}(t) \to u(t) = u_f(t)$ in E for each $t \in [0,1]$, $w\text{-}D^{\alpha-1} u_n(t) = w\text{-}D^{\alpha-1} u_{f_n}(t) \to w\text{-}D^{\alpha-1} u(t)$ in E for each $t \in [0,1]$, and $w\text{-}D^{\alpha} u_n = f_n$ $\sigma(P^1_E, L^\infty \otimes E^*)$-converges to $w\text{-}D^{\alpha} u = f$ and similarly $v_n(t) = u_{g_n}(t) \to v(t) = v_g(t)$ in E for each $t \in [0,1]$, $w\text{-}D^{\alpha-1} v_n(t) = w\text{-}D^{\alpha-1} u_{g_n}(t) \to w\text{-}D^{\alpha-1} v(t)$ in E for each $t \in [0,1]$, and $w\text{-}D^{\alpha} v_n = g_n$ $\sigma(P^1_E, L^\infty \otimes E^*)$-converges to $w\text{-}D^{\alpha} v = g$ with

$$g_n(t) \in F(t, u_{f_n}(t), w\text{-}D^{\alpha-1} u_{f_n}(t)) \ a.e.$$

From the above inclusion, we have

$$\langle h \otimes x^*, g_n(t) \rangle \le \delta^*(h \otimes x^*, F(t, u_{f_n}(t), w\text{-}D^{\alpha-1} u_{f_n}(t))) \ a.e.$$

where $h \in L^\infty([0,1], dt)$ and $x^* \in E^*$. By integrating we get

$$\int_0^1 \langle h \otimes x^*, g_n(t) \rangle dt \le \int_0^1 \delta^*(h \otimes x^*, F(t, u_{f_n}(t), w\text{-}D^{\alpha-1} u_{f_n}(t)))dt.$$

Whence we deduce that

$$\lim_{n\to\infty} \int_0^1 \langle h \otimes x^*, g_n(t) \rangle dt = \int_0^1 \langle h \otimes x^*, g(t) \rangle dt$$

$$\le \limsup_n \int_0^1 \delta^*(h \otimes x^*, F(t, u_{f_n}(t), w\text{-}D^{\alpha-1} u_{f_n}(t)))dt$$

$$\leq \int_0^1 \limsup_n \delta^*(h \otimes x^*, F(t, u_{f_n}(t), w\text{-}D^{\alpha-1}u_{f_n}(t)))dt$$

$$\leq \int_0^1 \delta^*(h \otimes x^*, F(t, u_f(t), w\text{-}D^{\alpha-1}u_f(t)))dt$$

so that, in particular, by taking $h = 1_A$ with $A \in \mathcal{L}_\lambda([0,1]$, we get

$$\int_A \langle x^*, g(t)\rangle dt \leq \int_A \delta^*(x^*, F(t, u_f(t), w\text{-}D^{\alpha-1}u_f(t)))dt$$

so that

$$\langle x^*, g(t)\rangle \leq \delta^*(x^*, F(t, u_f(t), w\text{-}D^{\alpha-1}u_f(t))) \ a.e.$$

By taking a dense sequence (e_k^*) with respect to the Mackey topology, we get

$$\langle e_k^*, g(t)\rangle \leq \delta^*(e_k^*, F(t, u_f(t), w\text{-}D^{\alpha-1}u_f(t))) \ a.e.$$

By virtue of ([17], Prop. III.35) we conclude that

$$g(t) \in F(t, u_f(t), w\text{-}D^{\alpha-1}u_f(t)) \ a.e.$$

thus proving that the graph of Φ is closed in $\mathcal{X} \times \mathcal{X}$. By the classical Kakutani-Ky Fan fixed point theorem, Φ admit a fixed point. This leads automatically to the existence and compactness property of the solutions set to (29). ■

We mention that some of the preceding results are valid when the second member F and the multifunction X take convex **weakly compact** values.

In that case, Theorem 4.3 is read as

Theorem 4.4. *Let $X : [0,1] \hookrightarrow E$ be a convex weakly compact valued Pettis-integrable multifunction. Let $F : [0,1] \times E_\sigma \times E_\sigma \hookrightarrow E$ be a convex weakly compact valued mapping satisfying*

(1) F is scalarly $\mathcal{L}_\lambda([0,1]) \otimes \mathcal{B}(E_\sigma) \otimes \mathcal{B}(E_\sigma)$-measurable, i.e., for each $x^* \in E^*$, the scalar function $\delta^*(x^*, F(.,.,.))$ is $\mathcal{L}_\lambda([0,1]) \otimes \mathcal{B}(E_\sigma) \otimes \mathcal{B}(E_\sigma)$-measurable,
(2) The mapping $F(t,.,.)$ is scalarly upper semicontinuous, i.e., for each $t \in [0,1]$ and for each $x^* \in E^*$, the scalar function $\delta^*(x^*, F(t,.,.))$ is sequentially upper semicontinuous on $E_\sigma \times E_\sigma$,
(3) $F(t,x,y) \subset X(t)$ for all $(t,x,y) \in [0,1] \times E \times E$.

Then the $W^{\alpha,1}_{P,E}([0,1])$*-solutions set to the (FDI)*

$$\begin{cases} w\text{-}D^{\alpha}u(t) \in F(t, u(t), w\text{-}D^{\alpha-1}u(t)),\ t \in [0,1] \\ u(0) = 0,\ w\text{-}D^{\alpha-1}u(0) = b \end{cases} \tag{30}$$

is weakly compact in the space $\mathcal{C}_E([0,1])$.

Similar results in the Bochner setting hold automatically.

5 Multivalued Integral Inclusion

The results stated in the preceding section lead naturally to the study of multivalued Pettis integral inclusion and the multivalued fractional Pettis integral inclusion. For simplicity we begin with an example.

Proposition 5.1. *Assume that* E *is a reflexive separable Banach space and* $F : [0,1] \times E \to E$ *is a mapping satisfying*

(i) F *is* $\mathcal{L}_\lambda([0,1]) \otimes \mathcal{B}(E)$*-measurable,*
(ii) For each $t \in [0,1]$, $F(t,.)$ *is sequentially weakly continuous,*
(iii) $||f(t,x)|| \leq h(t)$ *for all* $(t,x) \in 0,1] \times E$ *where* $h : [0,1] \to \mathbf{R}^+$ *is a positive integrable function.*

Let $H : [0,1] \times [0,1] \to \mathbf{R}$ *be a bounded Carathéodory mapping, that is,* $H(.,s)$ *is continuous for every fixed* $s \in [0,1]$ *and* $H(t,.)$ *is measurable for every fixed* $t \in [0,1]$.

Then the integral equation

$$x(t) = \int_0^1 H(t,s)F(s,x(s))ds,\ \forall t \in [0,1]$$

admits at least a solution $x \in \mathcal{C}_E([0,1])$.

Proof. Set $X(t) = h(t)\overline{B}_E$ for all $t \in [0,1]$. Since E is reflexive separable Banach space, X is a measurable integrably bounded convex weakly compact valued multimapping, and the set S^1_X of all measurable selections of X is convex $\sigma(L^1_E, L^\infty_{E^*})$-compact. It is not difficult to check that the set

$$\mathcal{Y} := \{u : [0,1] \to E : u(t) \in \int_0^1 H(t,s)X(s)ds, t \in [0,1];$$

$$||u(t) - u(\tau)|| \leq \int_0^1 |H(t,s) - H(\tau,s)|h(s)ds,\ \forall t,\tau \in [0,1]\}$$

is bounded convex equicontinuous weakly compact in $C_E([0,1])$. For each $x \in \mathcal{Y}$, let us set

$$\Phi(x) : t \mapsto \int_0^1 H(t,s)F(s,x(s))ds, \ \forall t \in [0,1].$$

It is clear that $\Phi(x) \in \mathcal{Y}$. We will show that the mapping $\Phi : \mathcal{Y} \to \mathcal{Y}$ is continuous on the convex weakly compact $\mathcal{Y}$. Let $x_n \in \mathcal{Y}$ with $x_n \to x$ weakly in $C_E([0,1])$. Then $x_n(t) \to x(t)$ weakly in E for each $t \in [0,1]$. By (ii), $F(t, x_n(t)) \to F(t, x(t))$ weakly in E for each $t \in [0,1]$. But $F(t, x_n(t)) \in X(t)$ for all $n \in \mathbf{N}$ and for all $t \in [0,1]$, so by the sequential $\sigma(L^1_E, L^\infty_{E^*})$-compactness of S^1_X, we may assume that $F(., x_n(.))$ $\sigma(L^1_E, L^\infty_{E^*})$-converges to $v \in S^1_X$ so that by identifying the limit, $F(., x_n(.))$ $\sigma(L^1_E, L^\infty_{E^*})$-converges to $F(., x(.))$. Indeed, for each $A \in \mathcal{L}_\lambda([0,1])$ and for each $x^* \in E^*$, we have,

$$\int_A \langle x^*, v(s)\rangle ds = \lim_n \int_A \langle x^*, F(s, x_n(s))\rangle ds = \int_A \langle x^*, F(s, x(s))\rangle ds$$

using (iii). Whence we get

$$\langle x^*, v(s)\rangle = \langle x^*, F(s, x(s))\rangle \ a.e.$$

for each $x^* \in E^*$. Since E is separable, $v(s) = F(s, x(s))$ a.e. Coming back to the definition of Φ, we note that

$$\Phi(x_n)(t) = \int_0^1 H(t,s)F(s,x_n(s))ds \to \int_0^1 H(t,s)F(s,x(s))ds = \Phi(x)(t)$$

in E_σ for each $t \in [0,1]$, that is, $\Phi(x_n)$ converges to $\Phi(x)$ weakly in $C_E([0,1])$ showing that $\Phi : \mathcal{Y} \to \mathcal{Y}$ is continuous on the convex weakly compact $\mathcal{Y}$. Hence Φ admits at least a fixed point $x = \Phi(x)$ that constitutes a solution of the integral equation under consideration. ■

In the vein of the above result, we provide a fairly general existence theorem in multivalued Pettis integral inclusion which has an independent interest and which leads to the existence of solutions for some differential inclusion and fractional differential inclusion with boundary conditions in the Pettis setting.

Theorem 5.1. *Let $H : [0,1] \times [0,1] \to \mathbf{R}$ be a mapping with the following properties*

(i) for each $t \in [0,1]$, $H(t,.)$ is Lebesgue-measurable on $[0,1]$,
(ii) for each $s \in [0,1]$, $H(.,s)$ is continuous on $[0,1]$,
(iii) there is a constant $M > 0$ such that $|H(t,s)| \le M$ for all $(t,s) \in [0,1] \times [0,1]$.

Let $X : [0,1] \hookrightarrow E$ be a convex weakly compact valued Pettis-integrable multimapping. Let $F : [0,1] \times E_\sigma \hookrightarrow E$ be a convex weakly compact valued multimapping satisfying

(1) F is scalarly $\mathcal{L}_\lambda([0,1]) \otimes \mathcal{B}(E_\sigma)$-measurable, i.e., for each $x^* \in E^*$, the scalar function $\delta^*(x^*, F(.,.))$ is $\mathcal{L}_\lambda([0,1]) \otimes \mathcal{B}(E_\sigma)$-measurable,
(2) The multimapping $F(t,.)$ is scalarly sequentially upper semicontinuous, i.e., for each $t \in [0,1]$ and for each $x^* \in E^*$, the scalar function $\delta^*(x^*, F(t,.))$ is sequentially upper semicontinuous on E_σ,
(3) $F(t,x) \subset X(t)$ for all $(t,x) \in [0,1] \times E$.

Then the multivalued Pettis integral inclusion

$$x(t) \in \int_0^1 H(t,s)F(s,x(s))ds, \ t \in [0,1]$$

admits at least a solution $x \in C_E([0,1])$.

Proof.

Step 1 Let us set

$$\mathcal{Y} := \{u : [0,1] \to E : u(t) \in \int_0^1 H(t,s)X(s)ds, \ t \in [0,1];$$

$$||u(t) - u(\tau)|| \leq \sup_{x^* \in \overline{B}_{E^*}} \int_0^1 |H(t,s) - H(\tau,s)||\delta^*(x^*, X(s))|ds, \ \forall t, \tau \in [0,1]\}$$

Then $\mathcal{Y}$ is bounded convex equicontinuous in $C_E([0,1])$. Further, $\mathcal{Y}(t)$ is relatively weakly compact, because it is included in the convex weakly compact $\int_0^1 H(t,s)X(s)ds$ for each $t \in [0,1]$ so that $\mathcal{Y}$ is relatively weakly compact in $C_E([0,1])$. We claim that $\mathcal{Y}$ is weakly compact in $C_E([0,1])$. Let (f_n) be a sequence in S^1_X. By the sequential $\sigma(P^1_E, L^\infty \otimes E^*)$-compactness of S^{Pe}_X, we may assume that (f_n) $\sigma(P^1_E, L^\infty \otimes E^*)$-converges to a Pettis integrable mapping $f \in S^{Pe}_X$. It follows that

$$u_f(t) = \text{w-}\lim_n u_{f_n}(t) = \text{w-}\lim_n \int_0^1 H(t,s)f_n(s)ds = \int_0^1 H(t,s)f(s)ds$$

for every $t \in [0,1]$ showing the weak compactness of $\mathcal{Y}$ in the space $C_E([0,1])$.

Step 2 For each $x \in \mathcal{Y}$ let us set

$$\Psi(x) := \{w_x : [0,1] \to E, \ w_x(t) = \int_0^1 H(t,s)v(s)ds, \ v \in S^1_{F(x)}, \ t \in [0,1]\}.$$

Then $\Psi(x)$ is a convex weakly compact subset of $\mathcal{Y}$. Let us check that the multimapping $\Psi : \mathcal{Y} \hookrightarrow \mathcal{Y}$ is upper semicontinuous, equivalently, its graph is sequentially weakly compact in $\mathcal{Y} \times \mathcal{Y}$. Let $y_n \in \Psi(x_n)$ with $x_n \to x$ and $y_n \to y$ weakly in $C_E([0,1])$. By $y_n \in \Psi(x_n)$, there exist $v_n \in S^1_{F(x_n)}$ such that $y_n(t) = \int_0^1 H(t,s)v_n(s)ds$ for all $t \in [0,1]$. As $F(t, x_n(t)) \subset X(t)$ for all $t \in [0,1]$ and for all $n \in \mathbf{N}$, by assumption (3), we have $v_n \in S_X^{Pe}$ for all $n \in \mathbf{N}$. By the sequential $\sigma(P_E^1, L^\infty \otimes E^*)$-compactness of S_X^{Pe}, we may assume that v_n $\sigma(P_E^1, L^\infty \otimes E^*)$-converges to a Pettis integrable mapping $v \in S_X^{Pe}$. As y_n weakly converges to y, we deduce that

$$y(t) = \text{w-}\lim_n y_n(t) = \text{w-}\lim_n \int_0^1 H(t,s)v_n(s)ds = \int_0^1 H(t,s)v(s)ds$$

for every $t \in [0,1]$. Now from the inclusion $v_n(t) \in F(t, x_n(t))$ we have, for each $x^* \in E^*$ and for each $A \in \mathcal{L}_\lambda([0,1])$

$$\langle 1_A(t)x^*, v_n(t)\rangle \leq \delta^*(1_A(t)x^*, F(t, x_n(t)))\ a.e.\ t \in [0,1]$$

so that by integrating

$$\int_A \langle x^*, v_n(t)\rangle dt \leq \int_A \delta^*(x^*, F(t, x_n(t)))dt.$$

Thus we have

$$\begin{aligned}
\lim_n \int_A \langle x^*, v_n(t)\rangle dt &= \int_A \langle x^*, v(t)\rangle dt \\
&\leq \limsup_n \int_A \delta^*(x^*, F(t, x_n(t)))dt \\
&\leq \int_A \limsup_n \delta^*(x^*, F(t, x_n(t)))dt \\
&\leq \int_A \delta^*(x^*, F(t, x(t)))dt.
\end{aligned}$$

Whence we get

$$\int_A \langle x^*, v(t)\rangle dt \leq \int_A \delta^*(x^*, F(t, x(t)))dt$$

for every $A \in \mathcal{L}_\lambda([0,1])$. Consequently

$$\langle x^*, v(t)\rangle \leq \delta^*(x^*, F(t, x(t)))\ a.e.$$

By taking a dense sequence (e_k^*) in E^* with respect to the Mackey topology $\tau(E^*, E)$, we get

$$\langle e_k^*, v(t)\rangle \le \delta^*(e_k^*, F(t, x(t))) \ a.e.$$

for all $k \in \mathbf{N}$, by ([17], Prop. III.35) we get finally

$$v(t) \in F(t, x(t)) \ a.e.$$

As $\mathcal{Y}$ is convex weakly compact in the space $C_E([0,1])$, it turns out that the $cwk(\mathcal{Y})$-valued $\Psi : \mathcal{Y} \hookrightarrow \mathcal{Y}$ is upper semicontinuous, then by the classical Kakutani-Ky Fan fixed point theorem, Ψ admits a fixed point, $x \in \Psi(x)$, that constitutes a solution of the multivalued integral inclusion under consideration. ■

The preceding result holds in the Bochner setting by assuming that X is convex weakly compact valued and integrably bounded multifunction.

Multivalued fractional Pettis integral

If $f : [0,1] \to E$ is a Pettis-integrable, by Lemmas 2.4 and 2.5, it is shown that the mapping $I^\alpha f : [0,1] \to E$ defined by

$$I^\alpha f(t) = \int_0^t \frac{(t-s)^{\alpha-1}}{\Gamma(\alpha)} f(s)ds, \ t \in [0,1]$$

belongs to $C_E([0,1])$ and satisfies some properties of convergence. In view of further application, we present now some properties of the multivalued fractional integral. Let $X : [0,1] \hookrightarrow E$ be a convex weakly compact valued Pettis-integrable mapping and let S_X^{Pe} be the set of all Pettis-integrable selection of X. The Aumann-Pettis fractional of X is defined by

$$I^\alpha X(t) = \{\int_0^t \frac{(t-s)^{\alpha-1}}{\Gamma(\alpha)} f(s)ds : f \in S_X^{Pe}\}, \ t \in [0,1].$$

It is clear that, for each $t \in [0,1]$, the multimapping $s \mapsto \frac{(t-s)^{\alpha-1}}{\Gamma(\alpha)} X(s)$ is convex weakly compact valued and Pettis-integrable so that the Aumann-Pettis fractional $I^\alpha X(t)$ of X is convex weakly compact valued, and by the classical Strassen theorem for the integral of convex weakly compact valued multimapping (see [17]) we have

$$\delta^*(x^*, I^\alpha X(t)) = \int_0^t \delta^*(x^*, \frac{(t-s)^{\alpha-1}}{\Gamma(\alpha)} X(s))ds$$

$$= \int_0^t \frac{(t-s)^{\alpha-1}}{\Gamma(\alpha)} \delta^*(x^*, X(s))ds = I^\alpha[\delta^*(x^*, X)](t)$$

for all $x^* \in E^*$ and for all $t \in [0, 1]$. The above considerations allow to define the Aumann-Bochner fractional of X when $X : [0, 1] \hookrightarrow E$ is a convex weakly compact valued integrably bounded multimapping. Similarly, the Aumann-Gelfand fractional of a convex weakly star compact valued Gelfand integrably bounded multimapping $X : [0, 1] \hookrightarrow E^*_s$, that is, X is scalarly integrable: $t \mapsto \delta^*(x, X(t))$ is integrable, for each $x \in E$, and $|X|$ is integrable. The Aumann-Gelfand fractional $I^\alpha X(t)$ of X is convex weakly star compact valued, and by the classical Strassen theorem for the integral of convex weakly compact valued multimapping (see [17]) we have

$$\delta^*(x, I^\alpha X(t)) = \int_0^t \delta^*(x, \frac{(t-s)^{\alpha-1}}{\Gamma(\alpha)} X(s))ds$$

$$= \int_0^t \frac{(t-s)^{\alpha-1}}{\Gamma(\alpha)} \delta^*(x, X(s))ds = I^\alpha[\delta^*(x, X)](t)$$

for all $x \in E$ and for all $t \in [0, 1]$.

The following establishes a sharp relationship between multivalued Aumann-Pettis integral and multivalued fractional Aumann-Pettis integral. Here $d_H(A, B)$ denotes the Hausdorff distance between two convex weakly compact subsets A and B in E.

Theorem 5.2. *Let $X : [0, 1] \hookrightarrow E$ be a convex weakly compact valued Pettis-integrable multimapping. Let $\alpha_n \in]1, 2]$ such that $\alpha_n \to 1$. Then, for every $t \in [0, 1]$, we have*

$$\lim_{n\to\infty} d_H(I^{\alpha_n} X(t), \int_0^t X(s)ds) = 0.$$

Proof. For every $t \in [0, 1]$, we have

$$\sup_{x^* \in \overline{B}_{E^*}} |\int_0^t \frac{(t-s)^{\alpha_n-1}}{\Gamma(\alpha_n)} \delta^*(x^*, X(s))ds - \int_0^t \delta^*(x^*, X(s)ds|$$

$$\leq \sup_{x^* \in \overline{B}_{E^*}} \int_0^t |\frac{(t-s)^{\alpha_n-1}}{\Gamma(\alpha_n)} - 1||\delta^*(x^*, X(s))|ds$$

As $|\frac{(t-s)^{\alpha_n-1}}{\Gamma(\alpha_n)} - 1| \to 0$ when $\alpha_n \to 1$, and $\{|\delta^*(x^*, X(s))| : x^* \in \overline{B}_{E^*}\}$ is uniformly integrable in $L^1_{\mathbf{R}}([0, 1])$, arguing as in the proof of Lemma 2.3, we conclude that

$$\lim_{n\to\infty} \sup_{x^* \in \overline{B}_{E^*}} \int_0^t |\frac{(t-s)^{\alpha_n-1}}{\Gamma(\alpha_n)} - 1||\delta^*(x^*, X(s))|ds = 0.$$

It follows that

$$d_H(I^{\alpha_n}X(t), \int_0^t X(s)ds) = \sup_{x^* \in \overline{B}_{E^*}} |\delta^*(x^*, I^{\alpha_n}X(t)) - \delta^*(x^*, \int_0^t X(s)ds)| \to 0$$

when $\alpha_n \to 1$. ■

When $X = f \in P^1_E([0,1])$ is a Pettis integrable mapping, the preceding result was stated in [21]. In Bochner case, Theorem 5.2 is read as

Theorem 5.3. *Let $X : [0,1] \hookrightarrow E$ be a convex weakly compact valued integrably bounded multimapping. Let $\alpha_n \in]1,2]$ such that $\alpha_n \to 1$. Then, every $t \in [0,1]$, we have*

$$\lim_{n\to\infty} d_H(I^{\alpha_n}X(t), \int_0^t X(s)ds) = 0.$$

We finish this section with some applications to a fractional integral inclusion.

Theorem 5.4. *Let E be a separable Banach space. Let $X : [0,1] \hookrightarrow E$ be a convex compact valued integrably bounded multimapping and let $F : [0,1] \times E \hookrightarrow E$ be a convex compact valued multimapping satisfying*

(1) F is scalarly $\mathcal{L}_\lambda([0,1]) \otimes \mathcal{B}(E)$-measurable, i.e., for each $x^* \in E^*$, the scalar function $\delta^*(x^*, F(.,.))$ is $\mathcal{L}_\lambda([0,1]) \otimes \mathcal{B}(E)$-measurable,
(2) The multimapping $F(t,.)$ is scalarly upper semicontinuous, that is, for each $t \in [0,1]$ and for each $x^* \in E^*$, the scalar function $\delta^*(x^*, F(t,.))$ is upper semicontinuous on E,
(3) $F(t,x) \subset X(t)$ for all $(t,x) \in [0,1] \times E$.

Then the multivalued fractional inclusion

$$x(t) \in \int_0^t \frac{|t-s|^{\alpha-1}}{\Gamma(\alpha)} F(s, x(s))ds, \ t \in [0,1]$$

admit at least a solution $x \in C_E([0,1])$.

Proof.

Step 1 Let us set

$$\mathcal{Y} := \{u : [0,1] \to E : u(0) = 0; \ u(t) \in \int_0^t \frac{|t-s|^{\alpha-1}}{\Gamma(\alpha)} X(s)ds, t \in [0,1];$$
$$||u(t) - u(\tau)|| \le \frac{|t-\tau|^{\alpha-1}}{\Gamma(\alpha)} \int_0^1 |X(s)|ds, \ \forall t, \tau \in [0,1]\}$$

Then $\mathcal{Y}$ is convex equicontinuous compact in $C_E([0,1])$ because the multivalued integral $\int_0^t \frac{|t-s|^{\alpha-1}}{\Gamma(\alpha)} X(s)ds$ is compact for each $t \in [0,1]$, see [11, 17].

Step 2 For each $x \in \mathcal{Y}$ let us set

$$\Psi(x) := \{w_x : [0,1] \to E,\ w_x(t) = \int_0^t \frac{|t-s|^{\alpha-1}}{\Gamma(\alpha)} v(s)ds,\ v \in S^1_{F(x)},\ t \in [0,1]\}.$$

Then $\Psi(x)$ is a convex compact subset of $\mathcal{Y}$. Let us check that the multimapping $\Psi : \mathcal{Y} \hookrightarrow \mathcal{Y}$ is upper semicontinuous, equivalently, its graph is sequentially closed in $\mathcal{Y} \times \mathcal{Y}$. Let $y_n \in \Psi(x_n)$ with $x_n \to x$ and $y_n \to y$ in $C_E([0,1])$. By $y_n \in \Psi(x_n)$, there exist $v_n \in S^1_{F(x_n)}$ such that $y_n(t) = \int_0^t \frac{|t-s|^{\alpha-1}}{\Gamma(\alpha)} v_n(s)ds$ for all $t \in [0,1]$. As $F(t, x_n(t)) \subset X(t)$ for all $t \in [0,1]$ and for all $n \in \mathbf{N}$ by assumption (3), we have $v_n \in S^1_X$ for all $n \in \mathbf{N}$. By the sequential $\sigma(L^1_E, L^\infty_{E^*})$-compactness of S^1_X, we may assume that v_n $\sigma(L^1_E, L^\infty_{E^*})$-converges to a Bochner integrable mapping $v \in S^1_X$. As y_n converges to y, we deduce that

$$y(t) = \lim_n y_n(t) = \lim_n \int_0^t \frac{|t-s|^{\alpha-1}}{\Gamma(\alpha)} v_n(s)ds = \int_0^t \frac{|t-s|^{\alpha-1}}{\Gamma(\alpha)} v(s)ds$$

for every $t \in [0,1]$. Now from the inclusion $v_n(t) \in F(t, x_n(t))$, we have, for each $x^* \in E^*$ and for each $A \in \mathcal{L}_\lambda([0,1])$

$$\langle 1_A(t)x^*, v_n(t)\rangle \leq \delta^*(1_A(t)x^*, F(t, x_n(t)))\ a.e.\ t \in [0,1]$$

so that by integrating

$$\int_A \langle x^*, v_n(t)\rangle dt \leq \int_A \delta^*(x^*, F(t, x_n(t)))dt.$$

Thus by (2) and (3) we have

$$\begin{aligned}\lim_n \int_A \langle x^*, v_n(t)\rangle dt &= \int_A \langle x^*, v(t)\rangle dt \\ &\leq \limsup_n \int_A \delta^*(x^*, F(t, x_n(t)))dt \\ &\leq \int_A \limsup_n \delta^*(x^*, F(t, x_n(t)))dt \\ &\leq \int_A \delta^*(x^*, F(t, x(t)))dt.\end{aligned}$$

Whence we get

$$\int_A \langle x^*, v(t)\rangle dt \leq \int_A \delta^*(x^*, F(t, x(t)))dt$$

for every $A \in \mathcal{L}_\lambda([0,1])$. Consequently

$$\langle x^*, v(t)\rangle \le \delta^*(x^*, F(t, x(t)))\ a.e.$$

By taking a dense sequence (e_k^*) in E^* with respect to the Mackey topology $\tau(E^*, E)$, we get

$$\langle e_k^*, v(t)\rangle \le \delta^*(e_k^*, F(t, x(t)))\ a.e.$$

for all $k \in \mathbf{N}$, by ([17], Prop. III.35) we get finally $v(t) \in F(t, x(t))$ *a.e.* so that $y \in \Psi(x)$. As $\mathcal{Y}$ is convex compact in the space $C_E([0,1])$, it turns out that the $ck(\mathcal{Y})$-valued $\Psi : \mathcal{Y} \hookrightarrow \mathcal{Y}$ is upper semicontinuous, then by the classical Kakutani-Ky Fan fixed point theorem, Ψ admits a fixed point, $x \in \Psi(x)$, that constitutes a solution of the multivalued fractional integral inclusion. ■

A Pettis variant of Theorem 5.4 is read as follows.

Theorem 5.5. *Let E be a separable Banach space. Let $X : [0,1] \hookrightarrow E$ be a convex weakly compact valued Pettis integrable multimapping and let $F : [0,1] \times E \hookrightarrow E$ be a convex weakly compact valued multimapping satisfying*

(1) F is scalarly $\mathcal{L}_\lambda([0,1]) \otimes \mathcal{B}(E)$-measurable, i.e., for each $x^* \in E^*$, the scalar function $\delta^*(x^*, F(.,.))$ is $\mathcal{L}_\lambda([0,1]) \otimes \mathcal{B}(E)$-measurable,
(2) The multimapping $F(t,.)$ is scalarly sequentially upper semicontinuous, i.e., for each $t \in [0,1]$ and for each $x^* \in E^*$, the scalar function $\delta^*(x^*, F(t,.))$ is sequentially upper semicontinuous on E_σ,
(3) $F(t,x) \subset X(t)$ for all $(t,x) \in [0,1] \times E$.

Then the multivalued fractional inclusion

$$x(t) \in \int_0^t \frac{|t-s|^{\alpha-1}}{\Gamma(\alpha)} F(s, x(s))ds,\ t \in [0,1]$$

admit at least a solution $x \in C_E([0,1])$.

Proof. The proof is omitted since it makes use of the machinery developed in the above section with appropriated modifications. ■

We provide two typical examples illustrating the techniques employed.

Proposition 5.2. *Assume that E is a separable reflexive Banach space and $F : [0,1] \times E \to E$ is a mapping satisfying*

(i) F is $\mathcal{L}_\lambda([0,1]) \otimes \mathcal{B}(E)$-measurable,
(ii) For each $t \in [0,1]$, $F(t,.)$ is sequentially weakly continuous,
(iii) $||f(t,x)|| \le h(t)$ for all $(t,x) \in [0,1] \times E$ where $h : [0,1] \to \mathbf{R}^+$ is a positive integrable function.

Then the integral equation

$$x(t) = \frac{t^{\alpha-1}}{\Gamma(\alpha)} \int_0^1 F(s, x(s))ds + \int_0^t \frac{|t-s|^{\alpha-1}}{\Gamma(\alpha)} F(s, x(s))ds, \ \forall t \in [0, 1]$$

admits at least a solution $x \in C_E([0,1])$. *The solutions set is weakly compact in* $C_E([0,1])$.

Proposition 5.3. *Assume that* E *is a separable Banach space and* $F : [0,1] \times E \times E \hookrightarrow E$ *is a convex compact valued multimapping satisfying*

(i) F *is* $\mathcal{L}_\lambda([0,1]) \otimes \mathcal{B}(E) \otimes \mathcal{B}(E)$*-measurable,*
(ii) For each $t \in [0,1]$, $F(t,.,.)$ *is scalarly upper semicontinuous, i.e., for each* $t \in [0,1]$ *and for each* $x^* \in E^*$*, the scalar function* $\delta^*(x^*, F(t,.,.))$ *is upper semicontinuous on* E,
(iii) $F(t,x,y) \subset X(t)$ *for all* $(t,x,y) \in [0,1] \times E \times E$ *where* $X : [0,1] \hookrightarrow E$ *is a convex compact valued integrably bounded multimapping.*

Let $A : C_E([0,1]) \to C_E([0,1])$ *be a continuous mapping. Then multivalued fractional integral inclusion*

$$x(t) \in \int_0^t \frac{|t-s|^{\alpha-1}}{\Gamma(\alpha)} F(s, x(s), Ax(s))ds, \ \forall t \in [0, 1]$$

admits at least a solution $x \in C_E([0,1])$. *The solutions set is compact in* $C_E([0,1])$.

We end this section with some variants related to the multivalued fractional integral of the level sets associated with a random fuzzy convex upper semicontinuous integrand $X : [0,1] \times E_\sigma \to [0,1]$. We first consider the fractional fuzzy differential inclusion

$$\begin{cases} w\text{-}D^\alpha x(t) \in L_\tau(X)(t), t \in [0,1], \tau \in]0,1] \\ x(0) = 0, w\text{-}D^{\alpha-1}x(0) = b \end{cases} \tag{31}$$

where $L_\tau(X)(s)$ is the level set of X.

Let us mention at first a useful result.

Lemma 5.1. *Let* E *be a separable Banach space. Let* $L :]0,1] \times [0,1] \hookrightarrow E$ *be a convex weakly compact valued multimapping satisfying*

(i) $L(\tau,.)$ *is scalarly is measurable on* $[0,1]$, *for every fixed* $\tau \in]0,1]$,
(ii) $L(.,s)$ *is scalarly left continuous on* $]0,1]$, *for every fixed* $s \in [0,1]$,
(iii) There is a convex weakly compact valued integrably bounded mapping $K : [0,1] \hookrightarrow E$ *such that* $L(\tau,s) \subset K(s)$ *for all* $\tau, s \in]0,1] \times [0,1]$.

Let $0 < \tau_1 < \tau_2 < \ldots < \tau_k \to \tau$ *and let* v_{τ_k} *be a measurable selection of* $L(\tau_k,.)$. *Then there is a subsequence* $(v_{\tau_k})_k$ *and an integrable mapping* $v : [0,1] \to E$ *such that* (v_{τ_k}) $\sigma(L^1_E, L^\infty_{E^*_s})$*-converges to* v *and such that* $v(s) \in L(\tau, s)$ *a.e.* $s \in [0,1]$.

Proof. We omit the proof since it follows from the technique given in the proof of Theorem 5.4. ∎

The following theorem summarizes the properties of the solutions set to the fractional fuzzy differential inclusion (31).

Theorem 5.6. *Let E be a reflexive separable Banach space. Let X be a random fuzzy convex upper semicontinuous integrand $X : [0,1] \times E_\sigma \to [0,1]$ with the following properties:*

(1) $\overline{\{x \in E_\sigma : X(s,x) > 0\}}$ *is weakly compact, for each $s \in [0,1]$,*
(2) $g := |L_{0^+}(X)| \in L^1$,
(3) For every fixed $s \in [0,1]$, the convex weakly compact valued multimapping $\tau \mapsto L_\tau(X)(s)$ is scalarly left continuous on $]0,1]$.

Then the following hold

(a)

$$\int_0^t \frac{|t-s|^{\alpha-1}}{\Gamma(\alpha)} L_\tau(X)(s)ds = \bigcap_{k\geq 1} \int_0^t \frac{|t-s|^{\alpha-1}}{\Gamma(\alpha)} L_{\tau_k}(X)(s)ds, \ t \in [0,1]$$

whenever $0 < \tau_1 < \tau_2 < \ldots < \tau_k \to \tau$.

(b) Let us consider

$$\mathcal{Y}_{\tau_k} = \{u_f : [0,1] \to E : u_f(t) = \frac{b}{\Gamma(\alpha)} t^{\alpha-1} + \frac{1}{\Gamma(\alpha)} \int_0^t (t-s)^{\alpha-1} f(s)ds,$$

$$\forall t \in [0,1]; \quad f \in S^1_{L_{\tau_k}(X)}\}.$$

Let $x_{\tau_k} \in \mathcal{Y}_{\tau_k}$. Then, up to an extracted subsequence, $(x_{\tau_k})_k$ converges weakly in $C_E([0,1])$ to a solution $x_\tau \in \mathcal{Y}_\tau$ where

$$\mathcal{Y}_\tau = \{u_f : [0,1] \to E : u_f(t) = \frac{b}{\Gamma(\alpha)} t^{\alpha-1} + \frac{1}{\Gamma(\alpha)} \int_0^t (t-s)^{\alpha-1} f(s)ds,$$

$$\forall t \in [0,1]; \quad f \in S^1_{L_\tau(X)}\}.$$

Consequently, for each $t \in [0,1]$,

$$w\text{-}\lim x_{\tau_k}(t) = x_\tau(t) \in \frac{b}{\Gamma(\alpha)} t^{\alpha-1} + \int_0^t \frac{|t-s|^{\alpha-1}}{\Gamma(\alpha)} L_\tau(X)(s)ds$$

$$= \frac{b}{\Gamma(\alpha)} t^{\alpha-1} + \bigcap_{k\geq 1} \int_0^t \frac{|t-s|^{\alpha-1}}{\Gamma(\alpha)} L_{\tau_k}(X)(s)ds.$$

Proof.

(a) Follows from Theorem 4.1 in [20] using the continuity property of the level sets.
(b) Using (2) it is already seen that $\mathcal{Y}_{\tau_k}$ is included in a convex weakly compact subset of $C_E([0,1])$. Hence by extracting subsequence we may assume that $(x_{\tau_k})_k$ converges weakly in $C_E([0,1])$ to a continuous mapping x. Consequently, for each $t \in [0,1]$, $w\text{-}\lim x_{\tau_k}(t) = x(t)$. But

$$x_{\tau_k}(t) = \frac{b}{\Gamma(\alpha)} t^{\alpha-1} + \int_0^t \frac{|t-s|^{\alpha-1}}{\Gamma(\alpha)} v_{\tau_k}(s)ds$$

where v_{τ_k} is a measurable selection of $L_{\tau_k}(X)$. Again by (2) and Lemma 5.1, there is a subsequence $(v_{\tau_k})_k$ and an integrable mapping $v_\tau : [0,1] \to E$ such that (v_{τ_k}) $\sigma(L^1_E, L^\infty_{E^*_s})$-converges to v_τ and such that $v_\tau(s) \in L_\tau(X)(s)$ a.e. Let

$$x_\tau(t) = \frac{b}{\Gamma(\alpha)} t^{\alpha-1} + \int_0^t \frac{|t-s|^{\alpha-1}}{\Gamma(\alpha)} v_\tau(s)ds.$$

Then, $x_\tau \in \mathcal{Y}_\tau$ and for each $t \in [0,1]$, we have

$$w\text{-}\lim_k x_{\tau_k}(t) = x_\tau(t) = \frac{b}{\Gamma(\alpha)} t^{\alpha-1} + \int_0^t \frac{|t-s|^{\alpha-1}}{\Gamma(\alpha)} v_\tau(s)ds \tag{32}$$

$$\in \frac{b}{\Gamma(\alpha)} t^{\alpha-1} + \int_0^t \frac{|t-s|^{\alpha-1}}{\Gamma(\alpha)} L_\tau(X)(s)ds \tag{33}$$

$$= \frac{b}{\Gamma(\alpha)} t^{\alpha-1} + \bigcap_{k\geq 1} \int_0^t \frac{|t-s|^{\alpha-1}}{\Gamma(\alpha)} L_{\tau_k}(X)(s)ds \tag{34}$$

$$= \frac{b}{\Gamma(\alpha)} t^{\alpha-1} + w\text{-}\lim_k \int_0^t \frac{|t-s|^{\alpha-1}}{\Gamma(\alpha)} L_{\tau_k}(X)(s)ds. \tag{35}$$

■

Actually (31) is a fractional differential inclusion with parameter $\tau \in]0,1]$ in the fuzzy setting, taking account of the specific properties of the solution set described in Theorem 5.6. It should be interesting to consider the fractional fuzzy differential inclusion

$$\begin{cases} w\text{-}D^\alpha x_\tau(t) \in L_\tau(X)(t) + F(t, x_\tau(t)), t \in [0,1], \tau \in]0,1] \\ x_\tau(0) = 0, w\text{-}D^{\alpha-1} x_\tau(0) = b \end{cases} \tag{36}$$

where $F : [0,1] \times E_\sigma \to E_\sigma$ satisfies

(1) F is scalarly $\mathcal{L}_\lambda([0,1]) \otimes \mathcal{B}(E)$-measurable, i.e., for each $x^* \in E^*$, the scalar function $\langle x^*, F(.,.)\rangle$ is $\mathcal{L}_\lambda([0,1]) \otimes \mathcal{B}(E)$-measurable,

(2) The mapping $F(t,.)$ is scalarly sequentially continuous, i.e., for each $t \in [0,1]$ and for each $x^* \in E^*$, the scalar function $\langle x^*, F(t,.)\rangle$ is sequentially continuous on E_σ,

(3) $||F(t,x)|| \leq h(t)$ for all $(t,x) \in [0,1] \times E_\sigma$.

Then the family of multimappings $G_\tau(t,x) := L_\tau(X)(t) + F(t,x)$ enjoys good properties of measurability and upper semicontinuity properties with $|G_\tau(t,x)| \leq k(t)$ for all $\tau \in]0,1]$ and for all $t \in [0,1]$ where $k = g + h$ is a positive integrable function. Now, for every $\tau \in]0,1]$, by the above considerations, it is clear that the integral inclusion

$$x_\tau(t) \in \frac{b}{\Gamma(\alpha)} t^{\alpha-1} + \int_0^t \frac{|t-s|^{\alpha-1}}{\Gamma(\alpha)} G_\tau(s, x_\tau(s))ds$$

admits at least a solution $x_\tau \in C_E([0,1])$. Namely $x_\tau \in \mathcal{Z}_\tau$ where

$$\mathcal{Z}_\tau = \{u_f : [0,1] \to E : u_f(t) = \frac{b}{\Gamma(\alpha)} t^{\alpha-1} + \frac{1}{\Gamma(\alpha)} \int_0^t (t-s)^{\alpha-1} f(s)ds,$$
$$\forall t \in [0,1]; \quad f(s) \in L_\tau(X)(s) + F(s, u_f(s)), \ a.e. \ s \in [0,1]\}.$$

Now let $0 < \tau_1 < \tau_2 < \ldots \tau_k \to \tau$ and let $x_{\tau_k} \in \mathcal{Z}_{\tau_k}$. We have for each $t \in [0,1]$

$$x_{\tau_k}(t) = \frac{b}{\Gamma(\alpha)} t^{\alpha-1} + \int_0^t \frac{|t-s|^{\alpha-1}}{\Gamma(\alpha)} v_{\tau_k}(s)ds$$

with $v_{\tau_k}(s) \in L_{\tau_k}(X)(s) + F(s, x_{\tau_k}(s))$, $a.e.\ s \in [0,1]$. Applying mutatis mutandis the above argument, it is already seen that $\mathcal{Z}_{\tau_k}$ is included a convex weakly compact subset of $C_E([0,1])$. Hence by extracting subsequence we may assume that $(x_{\tau_k})_k$ converges weakly in $C_E([0,1])$ to a continuous mapping x. Consequently, for each $t \in [0,1]$, $w\text{-}\lim x_{\tau_k}(t) = x(t)$. Again by (2) and arguing as Lemma 5.1, there is a subsequence $(v_{\tau_k})_k$ and an integrable mapping $v_\tau : [0,1] \to E$ such that (v_{τ_k}) $\sigma(L^1_E, L^\infty_{E^*_s})$-converges to v_τ and such that $v_\tau(s) \in L_\tau(X)(s) + F(s, x(s))$ a.e. Let

$$x_\tau(t) = \frac{b}{\Gamma(\alpha)} t^{\alpha-1} + \int_0^t \frac{|t-s|^{\alpha-1}}{\Gamma(\alpha)} v_\tau(s)ds$$

Then $x_\tau \in \mathcal{Z}_\tau$ and, for each $t \in [0,1]$, we have,

$$w\text{-}\lim_k x_{\tau_k}(t) = x_\tau(t) = \frac{b}{\Gamma(\alpha)} t^{\alpha-1} + \int_0^t \frac{|t-s|^{\alpha-1}}{\Gamma(\alpha)} v_\tau(s)ds \tag{37}$$

$$\in \frac{b}{\Gamma(\alpha)} t^{\alpha-1} + \int_0^t \frac{|t-s|^{\alpha-1}}{\Gamma(\alpha)} [L_\tau(X)(s) + F(s, x_\tau(s))]\, ds \tag{38}$$

$$= \frac{b}{\Gamma(\alpha)} t^{\alpha-1} + \int_0^t \frac{|t-s|^{\alpha-1}}{\Gamma(\alpha)} F(s, x_\tau(s))ds + \int_0^t \frac{|t-s|^{\alpha-1}}{\Gamma(\alpha)} L_\tau(X)(s)ds \tag{39}$$

$$= \frac{b}{\Gamma(\alpha)} t^{\alpha-1} + \int_0^t \frac{|t-s|^{\alpha-1}}{\Gamma(\alpha)} F(s, x_\tau(s))ds + \bigcap_{k\geq 1} \int_0^t \frac{|t-s|^{\alpha-1}}{\Gamma(\alpha)} L_{\tau_k}(X)(s)ds$$

$$= \frac{b}{\Gamma(\alpha)} t^{\alpha-1} + \int_0^t \frac{|t-s|^{\alpha-1}}{\Gamma(\alpha)} F(s, x_\tau(s))ds + w\text{-}\lim_k \int_0^t \frac{|t-s|^{\alpha-1}}{\Gamma(\alpha)} L_{\tau_k}(X)(s)ds. \tag{40}$$

In this regard one may consider the following example of fractional fuzzy differential inclusion

$$\begin{cases} w\text{-}D^\alpha x_\tau(t) \in \theta(t, x_\tau(t)) L_\tau(X)(t), t \in [0,1], \tau \in]0,1] \\ x_\tau(0) = 0, w\text{-}D^{\alpha-1} x_\tau(0) = b \end{cases} \tag{41}$$

where the coefficient $\theta : [0,1] \times E_\sigma \to [0,1]$ satisfies

(1) θ is $\mathcal{L}_\lambda([0,1]) \otimes \mathcal{B}(E)$-measurable,
(2) The mapping $\theta(t,.)$ is sequentially continuous on E_σ

in order to show the difference with (31) and (36). It is clear that the multimapping $G_\tau(t,x) := \theta(t,x) L_\tau(X)(t)$ enjoys good measurability and upper semicontinuity properties.

These considerations allow us to present a **variational** formulation on the solutions set of a family of fractional differential inclusions in a separable Banach space of the form

$$\begin{cases} w\text{-}D^\alpha x_\tau(t) \in G_\tau(t, x_\tau(t)), t \in [0,1], \tau \in]0,1] \\ x_\tau(0) = 0, w\text{-}D^{\alpha-1} x_\tau(0) = b \end{cases} \tag{42}$$

where $G_\tau(t,x)$ is a family of upper semicontinuous multimappings.

Theorem 5.7. *Let E be a separable Banach space. Let $X : [0,1] \hookrightarrow E$ be a convex weakly compact valued integrably bounded multimapping and let $G :]0,1] \times [0,1] \times E \hookrightarrow E$ be a convex weakly compact valued multimapping satisfying the conditions*

(1) For each fixed $\tau \in]0,1]$, $G_\tau(.,.)$ is scalarly $\mathcal{L}_\lambda([0,1]) \otimes \mathcal{B}(E)$-measurable, i.e., for each $x^* \in E^*$, the scalar function $\delta^*(x^*, G_\tau(.,.))$ is $\mathcal{L}_\lambda([0,1]) \otimes \mathcal{B}(E)$-measurable,
(2) For any fixed $t \in [0,1]$, $(\tau, x) \mapsto G_\tau(t,x)$ is scalarly sequentially upper semicontinuous on $]0,1] \times E_\sigma$, in the sense that, for $x^* \in E^*$, for any sequence $\tau_k \to \tau$ in $]0,1]$, for any sequence x_k weakly converging to x in E,

$$\limsup_k \delta^*(x^*, G_{\tau_k}(t, x_k)) \leq \delta^*(x^*, G_\tau(t,x))$$

(3) $G_\tau(t,x) \subset X(t)$ *for all* $(\tau,t,x) \in]0,1] \times [0,1] \times E$.

Then the following hold

(a) For each $k \in \mathbf{N}$, *the solutions set* $\mathcal{Z}_{\tau_k}$ *of the fractional inclusion*

$$\begin{cases} D^\alpha x_{\tau_k}(t) \in G_{\tau_k}(t, x_{\tau_k}(t)), t \in [0,1], \\ x_{\tau_k}(0) = 0, D^{\alpha-1}x_{\tau_k}(0) = b \end{cases}$$

is weakly compact in $C_E([0,1])$.

(b) Assume that $\tau_k \to \tau$ *in* $]0,1]$, *then* $w\text{-}ls\mathcal{Z}_{\tau_k} \subset \mathcal{Z}_\tau$.

Proof. With the above notations and hypotheses, the solutions set $\mathcal{Z}_{\tau_k}$ is given by

$$\mathcal{Z}_{\tau_k} = \{u_f : [0,1] \to E : u_f(t) = \frac{b}{\Gamma(\alpha)}t^{\alpha-1} + \frac{1}{\Gamma(\alpha)}\int_0^t (t-s)^{\alpha-1} f(s)ds,$$

$$\forall t \in [0,1]; \quad f(t) \in G_{\tau_k}(t, u_f(t))\}.$$

It is already seen that $\mathcal{Z}_{\tau_k}$ is a weakly compact subset of $C_E([0,1])$.

Now let $x \in w\text{-}ls\mathcal{Z}_{\tau_k}$. That means that there is $x_{\tau_k} \in \mathcal{Z}_{\tau_k}$ such that x_{τ_k} weakly converges to x in $C_E([0,1])$. We have

$$x_{\tau_k}(t) = \frac{b}{\Gamma(\alpha)}t^{\alpha-1} + \frac{1}{\Gamma(\alpha)}\int_0^t (t-s)^{\alpha-1} v_{\tau_k}(s)ds$$

with $v_{\tau_k}(s) \in G_{\tau_k}(s, x_{\tau_k}(s))$ a.e. By (1)–(3) and arguing as in the proof of Step 2 in Theorem 5.1, there are a subsequence $(v_{\tau_k})_k$ and an integrable mapping $v_\tau : [0,1] \to E$ such that (v_{τ_k}) $\sigma(L^1_E, L^\infty_{E^*_s})$-converges to v_τ and such that $v_\tau(s) \in G_\tau(s, x_\tau(s))$ a.e. Let

$$x_\tau(t) = \frac{b}{\Gamma(\alpha)}t^{\alpha-1} + \int_0^t \frac{|t-s|^{\alpha-1}}{\Gamma(\alpha)} v_\tau(s)ds$$

Then $x_\tau \in \mathcal{Z}_\tau$ and, for each $t \in [0,1]$, we have,

$$x(t) = w\text{-}\lim_k x_{\tau_k}(t) = x_\tau(t) = \frac{b}{\Gamma(\alpha)}t^{\alpha-1} + \int_0^t \frac{|t-s|^{\alpha-1}}{\Gamma(\alpha)} v_\tau(s)ds$$

showing that $w\text{-}ls\mathcal{Z}_{\tau_k} \subset \mathcal{Z}_\tau$. ∎

As a consequence of Theorem 5.7, we get a **stability** result concerning the inclusion

$$\begin{cases} w\text{-}D^\alpha x_\tau(t) \in G_\iota(t, x_\tau(t)), t \in [0,1], \tau \in]0,1] \\ x_\tau(0) = 0, w\text{-}D^{\alpha-1}x_\tau(0) = b \end{cases} \tag{43}$$

Several variants of Theorem 5.7 are available using the tools developed above. We omit the details. We refer to [1, 32] where several related results and references can be found.

In the next section, we study fractional differential equation with Young measure control.

6 Relaxation and Control Problem

Let E be a separable Banach space and let Z be a compact metric space. By $\mathcal{M}^1_+(Z)$ we denote the space of all probability Radon measures on Z. It is well-known that $\mathcal{M}^1_+(Z)$ is a compact metrizable space with respect to the narrow topology. We also denote by $\mathcal{Y}([0,1], \mathcal{M}^1_+(Z))$ the space of all Young measures defined on $[0,1]$ endowed with the stable topology. A sequence (ν^n) in $\mathcal{Y}([0,1], \mathcal{M}^1_+(Z))$ stably converges to $\nu \in \mathcal{Y}([0,1], \mathcal{M}^1_+(Z))$ if

$$\lim_{n\to\infty} \int_0^1 [\int_Z h_t(z) d\nu^n_t(z)] dt = \int_0^1 [\int_Z h_t(z) d\nu_t(z)] dt$$

for all $h \in L^1_{\mathcal{C}(Z)}([0,1])$, here $\mathcal{C}(Z)$ denotes the space of all continuous real valued functions defined on Z endowed with the norm of uniform convergence.

Let Δ be a measurable mapping defined on $[0,1]$ with nonempty compact values in Z and let $f : E \times E \times Z \to E$ be a function satisfying the following conditions

(i) f is continuous on $[0,1] \times E \times E \times Z$,
(ii) $f(t,x,y,z) \in K$ for all $(t,x,y,z) \in [0,1] \times E \times E \times Z$, where K is a convex compact subset of E,
(iii) there exist $l_1, l_2 \in \mathbf{R}^+$ with $(l_1 + l_2) < 1/2$ such that $||f(t,x_1,y_1,z) - f(t,x_2,y_2,z)|| \leq l_1||x_1 - x_2|| + l_2||y_1 - y_2|||$, for all (t,x_1,y_1,z) and $(t,x_2,y_2,z) \in [0,1] \times E \times E \times Z$.

We consider the following control problem governed by a (FDE) with original controls

$$\begin{cases} w\text{-}D^\alpha u_\zeta(t) = f(t, u_\zeta(t), w\text{-}D^{\alpha-1} u_\zeta(t), \zeta(t)) \\ u_\zeta(0) = 0, \ w\text{-}D^{\alpha-1} u_\zeta(0) = b \end{cases} \tag{44}$$

where ζ belongs to the set $\mathcal{Z} := S^1_\Delta$ of all original controls, which means that ζ is a Lebesgue-measurable selection of Δ. The existence of $W^{\alpha,1}_{B,E}([0,1])$ solution for (44) is guaranteed by Theorem 4.2 using conditions (i) and (ii). Let us check the uniqueness. Let u_{ζ_1}, u_{ζ_2} be two $W^{\alpha,1}_{B,E}([0,1])$-solutions to (44). Then, using (iii), for each $t \in [0, 1]$, we have

$$(^*) \quad ||w\text{-}D^\alpha u_{\zeta_1}(t) - w\text{-}D^\alpha u_{\zeta_2}(t)\|$$
$$= ||f(t, u_{\zeta_1}(t), w\text{-}D^{\alpha-1}u_{\zeta_1}(t), \zeta_1(t)) - f(t, u_{\zeta_2}(t), w\text{-}D^{\alpha-1}u_{\zeta_2}(t), \zeta_2(t))||$$
$$\leq l_1||u_{\zeta_1}(t) - u_{\zeta_2}(t)|| + l_2||w\text{-}D^{\alpha-1}u_{\zeta_1}(t) - w\text{-}D^{\alpha-1}u_{\zeta_2}(t)||.$$

On the other hand, it follows from Lemma 4.3, we have

$$u_\zeta(t) = \frac{1}{\Gamma(\alpha)} \int_0^t (t-s)^{\alpha-1} w\text{-}D^\alpha u_\zeta(s)ds + \frac{b}{\Gamma(\alpha)} t^{\alpha-1}, t \in [0, 1]$$

$$w\text{-}D^{\alpha-1}u_\zeta(t) = \int_0^t w\text{-}D^\alpha u_\zeta(s)ds + b$$

for all $t \in [0, 1]$ so that

$$||u_{\zeta_1}(t) - u_{\zeta_2}(t)|| \leq ||w\text{-}D^\alpha u_{\zeta_1} - w\text{-}D^\alpha u_{\zeta 2}||_{L^1_E([0,1])} \qquad (^{**})$$

$$||w\text{-}D^{\alpha-1}u_{\zeta_1}(t) - w\text{-}D^{\alpha-1}u_{\zeta_2}(t)| \leq ||w\text{-}D^\alpha u_{\zeta_1} - w\text{-}D^\alpha u_{\zeta 2}||_{L^1_E([0,1])} \qquad (^{***})$$

for every $t \in [0, 1]$. From (**)–(***) and (ii) we deduce that

$$\left\| w\text{-}D^\alpha u_{\zeta_1} - w\text{-}D^\alpha u_{\zeta 2} \right\|_{L^1_E([0,1])} \leq (l_1 + l_2)||w\text{-}D^\alpha u_{\zeta_1} - w\text{-}D^\alpha u_{\zeta_2}||_{L^1_E([0,1])}$$
$$< 1/2 \left\| w\text{-}D^\alpha u_{\zeta_1} - w\text{-}D^\alpha u_{\zeta_2} \right\|_{L^1_E([0,1])}$$

which ensures that $w\text{-}D^\alpha u_{\zeta_1} = w\text{-}D^\alpha u_{\zeta_2}$. Hence by (**) we get $u_{\zeta_1} = u_{\zeta_2}$.

Next we consider a relaxation problem governed by a (FDE) with Young measure controls

$$\begin{cases} w\text{-}D^\alpha u_\nu(t) = \int_Z f\left(u_\nu(t), w\text{-}D^{\alpha-1}u_\nu(t), z\right) \nu_t(dz) \\ u_\nu(0) = 0, \ w\text{-}D^{\alpha-1}u_\nu(0) = b \end{cases} \qquad (45)$$

where ν belongs to the set $\mathcal{R} := S_\Sigma$ of all relaxed (alias Young measure) controls, which means that ν is a Lebesgue-measurable selection of the mapping Σ defined by

$$\Sigma(t) := \left\{\sigma \in \mathcal{M}^1_+(Z) : \sigma(\Delta(t)) = 1\right\}$$

for all $t \in [0, 1]$. We note that, for each $\nu \in \mathcal{R}$, the existence and uniqueness of $W^{\alpha,1}_{B,E}([0, 1])$-solution to (45) again follows from the above remark, because the function

$$\tilde{f}(t,x,y,\nu) = \int_Z f(t,x,y,z)\nu(dz)$$

$(t,x,y,\nu) \in [0,1]\times E\times E\times \mathcal{M}^1_+(Z)$, inherits the properties of the function f. Namely we have

Theorem 6.1. *Let (i) – (iii) hold. Then, for each $\nu \in \mathcal{R}$, the (FDE) (45) has a unique solution $u_\nu \in W^{\alpha,1}_{B,E}([0,1])$. Further,*

$$u_\nu(t) = \frac{1}{\Gamma(\alpha)}\int_0^t (t-s)^{\alpha-1})w\text{-}D^\alpha u_\nu(s)ds + \frac{b}{\Gamma(\alpha)}t^{\alpha-1}, t \in [0,1]$$

$$w\text{-}D^{\alpha-1}u_\nu(t) = \int_0^t w\text{-}D^\alpha u_\nu(s)ds + b$$

and

$$w\text{-}D^\alpha u_\nu(t) = \int_Z f(t,u_\nu(t),w\text{-}D^{\alpha-1}u_\nu(t),z)\nu_t(dz)$$

for all $t \in [0,1]$.

Now we will state some topological properties of the $W^{\alpha,1}_{B,E}([0,1])$-solutions set to the FDE (45).

Theorem 6.2. *Let (i) – (iii) hold. Then the set $(\mathcal{S}_\mathcal{R})$ of all $W^{\alpha,1}_{B,E}([0,1])$-solutions to (45) is compact in $\mathcal{C}_E([0,1])$. Moreover the $W^{\alpha,1}_{B,E}([0,1])$-solutions set $(\mathcal{S}_\mathcal{Z})$ to (44) is dense in $(\mathcal{S}_\mathcal{R})$ with respect to the topology of uniform convergence.*

Proof. The proof is divided by two steps.

Step 1. Let (ν^n) be a sequence in $\mathcal{R} \equiv S_\Sigma$ which stably converges to $\nu^\infty \in \mathcal{R}$, and, for each $n \in \mathbf{N}\cup\{\infty\}$, let u_{ν^n} be the unique solution to the (FDE)

$$\begin{cases} w\text{-}D^\alpha u_{\nu^n}(t) = \int_Z f\left(t,u_{\nu^n}(t),w\text{-}D^{\alpha-1}u_{\nu^n}(t),z\right)\nu^n_t(dz) \\ u_\nu(0) = 0,\ w\text{-}D^{\alpha-1}u_\nu(0) = b \end{cases}$$

From Theorem 6.1 we have

$$u_{\nu^n}(t) = \frac{1}{\Gamma(\alpha)}\int_0^t (t-s)^{\alpha-1})w\text{-}D^\alpha u_{\nu^n}(s)ds + \frac{b}{\Gamma(\alpha)}t^{\alpha-1}, t \in [0,1] \tag{46}$$

$$w\text{-}D^{\alpha-1}u_{\nu^n}(t) = \int_0^t D^\alpha u_{\nu^n}(s)ds + b \tag{47}$$

$$w\text{-}D^\alpha u_{\nu^n}(t) = \int_Z f\left(t,u_{\nu^n}(t),w\text{-}D^{\alpha-1}u_{\nu^n}(t),z\right)\nu^n_t(dz) \in \rho\overline{B}_E \tag{48}$$

for all $t \in [0,1]$. From the above facts and Theorem 4.3, (u_{ν^n}) is relatively compact in $\mathcal{C}_E([0,1])$. Applying the techniques developed in the proof of Theorem 4.2, we may assert that

(a) u_{ν^n} pointwise converges on $[0,1]$ to $u_\infty \in W^{\alpha,1}_{B,E}([0,1])$,
(b) $w\text{-}D^{\alpha-1}u_{\nu^n}$ pointwise converges to $w\text{-}D^{\alpha-1}u_\infty$,
(c) $w\text{-}D^{\alpha}u_{\nu^n}$ $\sigma(L^1_E, L^\infty_E)$-converges to $w\text{-}D^{\alpha}u_\infty$,
with
(d) $u_\infty(0) = 0,\ w\text{-}D^{\alpha-1}u_\infty(0) = b$.

Step 2. We claim that u_∞ coincides with the $W^{\alpha,1}_{B,E}([0,1])$-solution u_{ν^∞} associated with $\nu^\infty \in \mathcal{R}$ to the (FDE)

$$\begin{cases} w\text{-}D^{\alpha}u_{\nu^\infty}(t) = \int_Z f\left(t, u_{\nu^\infty}(t), w\text{-}D^{\alpha-1}u_{\nu^\infty}(t), z\right)\nu^\infty_t(dz) \\ u_{\nu^\infty}(0) = 0,\ w\text{-}D^{\alpha-1}u_{\nu^\infty}(0) = b \end{cases}$$

Indeed, multiplying scalarly the Eq. (48)

$$w\text{-}D^{\alpha}u_{\nu^n}(t) = \int_Z f\left(t, u_{\nu^n}(t), w\text{-}D^{\alpha-1}u_{\nu^n}(t), z\right)\nu^n_t(dz)$$

by $h \in L^\infty_E([0,1])$ and integrating on $[0,1]$ we get

$$\int_0^1 \langle h(t), w\text{-}D^{\alpha}u_{\nu^n}(t)\rangle\, dt = \int_0^1 \left\langle h(t), \int_Z f\left(t, u_{\nu^n}(t), w\text{-}D^{\alpha-1}u_{\nu^n}(t), z\right)\nu^n_t(dz)\right\rangle dt$$

so that

$$\begin{aligned} \lim_{n\to\infty}\int_0^1 \langle h(t), w\text{-}D^{\alpha}u_{\nu^n}(t)\rangle dt &= \int_0^1 \langle h(t), w\text{-}D^{\alpha}u_\infty(t)\rangle dt \\ = \lim_{n\to\infty}\int_0^1 &\left\langle h(t), \int_Z f\left(t, u_{\nu^n}(t), w\text{-}D^{\alpha-1}u_{\nu^n}(t), z\right)\nu^n_t(dz)\right\rangle dt \\ = \lim_{n\to\infty}\int_0^1 &\left[\int_Z \langle h(t), f(t, u_{\nu^n}(t), w\text{-}D^{\alpha-1}u_{\nu^n}(t), z)\rangle \nu^n_t(dz)\right]dt. \end{aligned} \tag{49}$$

Using (a), (b), the stable convergence $\nu^n \to \nu^\infty$ and the fiber product theorem for Young measures ([18], Theorem 3.3.3), we note that the fiber product $\delta_{u_{\nu^n}} \otimes \delta_{w\text{-}D^{\alpha-1}u_{\nu^n}} \otimes \nu^n$ stably converges to $\delta_{u_\infty} \otimes \delta_{w\text{-}D^{\alpha-1}u_\infty} \otimes \nu^\infty$. Further the Carathéodory integrand $H(t,x,y,z) := \langle h(t), f(t,x,y,z)\rangle$ satisfies the following estimate

$$|H(t,x,y,z)| = |\langle h(t), f(t,x,y,z)\rangle| \le \rho||h(t)||$$

with $t \mapsto \rho||h(t)|| \in L^1_{\mathbf{R}^+}([0,1])$. From the above fact, we deduce that

$$\lim_{n\to\infty} \int_0^1 [\int_Z H(t, u_{\nu^n}(t), w\text{-}D^\alpha u_{\nu^n}(t), z)\nu_t^n(dz)]dt$$
$$= \int_0^1 [\int_Z H(t, u_\infty(t), w\text{-}D^\alpha u_\infty(t), z)\nu_t^\infty(dz)]dt \tag{50}$$

Then from (49) and (50) we get

$$\int_0^1 \langle h(t), w\text{-}D^\alpha u_\infty(t)\rangle dt = \int_0^1 [\int_Z H(t, u_\infty(t), w\text{-}D^\alpha u_\infty(t), z)\nu_t^\infty(dz)]dt$$
$$= \int_0^1 \langle h(t), \int_Z f(t, u_\infty(t), w\text{-}D^\alpha u_\infty(t), z)\nu_t^\infty(dz)\rangle dt.$$

It follows that

$$w\text{-}D^\alpha u_\infty(t) = \int_Z f(t, u_\infty(t), w\text{-}D^\alpha u_\infty(t), z)\nu_t^\infty(dz), \ \ a.e.$$

Finally, by the uniqueness of the solutions, we conclude that $u_\infty = u_{\nu^\infty}$. This proves the first part of the theorem, while the second part follows by continuity and density since $\mathcal{Z}$ is dense in $\mathcal{R}$ with respect to the stable topology ([18], Lemma 7.1.1). ■

Now let $L : [0,1] \times E \times E \times Z \to \mathbf{R}$ be a bounded continuous mapping

$$|L(t,x,y,z)| \leq \eta, \ \ \forall (t,x,y,z) \in [0,1] \times E \times E \times Z$$

for some positive constant η. We will consider the following Bolza control problem.

Theorem 6.3. *Let us consider the problems*

$$\inf_{\zeta\in\mathcal{Z}} \int_0^1 L(t, u_\zeta(t), w\text{-}D^{\alpha-1}u_\zeta(t), \zeta(t))dt \tag{$\mathcal{P}_\mathcal{Z}$}$$

$$\inf_{\nu\in\mathcal{R}} \int_0^1 [\int_Z L(t, u_\nu(t), w\text{-}D^{\alpha-1}u_\nu(t), z)d\nu_t(z)]dt \tag{$\mathcal{P}_\mathcal{R}$}$$

where u_ζ is the $W^{\alpha,1}_{B,E}([0,1])$-solution associated with $\zeta \in \mathcal{Z}$ to

$$\begin{cases} w\text{-}D^\alpha u_\zeta(t) = f\left(t, u_\zeta(t), w\text{-}D^{\alpha-1}u_\zeta(t), \zeta(t)\right) \\ u_\zeta(0) = 0, \ w\text{-}D^{\alpha-1}u_\zeta(0) = b. \end{cases}$$

and u_ν is the $W^{\alpha,1}_{B,E}([0,1])$-solution associated with $\nu \in \mathcal{R}$ to

$$\begin{cases} w\text{-}D^\alpha u_\nu(t) = \int_Z f\left(t, u_\nu(t), w\text{-}D^{\alpha-1}u_\nu(t), z\right) \nu_t(dz), \\ u_\nu(0) = 0,\ w\text{-}D^{\alpha-1}u_\nu(0) = b. \end{cases}$$

Then we have

$$\inf_{\zeta\in\mathcal{Z}} \int_0^1 L(t, u_\zeta(t), w\text{-}D^{\alpha-1}u_\zeta(t), \zeta(t))dt = \inf_{\nu\in\mathcal{R}} \int_0^1 [\int_Z L(t, u_\nu(t), w\text{-}D^{\alpha-1}u_\nu(t), z)d\nu_t(z)]dt.$$

Proof. Let $\nu \in \mathcal{R}$. As $\mathcal{Z}$ is dense in $\mathcal{R}$ with respect to the stable topology, there exists a sequence (ζ_n) in $\mathcal{Z}$ such that the associated Young measures (δ_{ζ_n}) stably converges to ν. Let u_{ζ_n} be the unique solution to

$$\begin{cases} w\text{-}D^\alpha u_{\zeta_n}(t) = f\left(t, u_{\zeta_n}(t), w\text{-}D^{\alpha-1}u_{\zeta_n}(t), \zeta_n(t)\right) \\ u_{\zeta_n}(0) = 0,\ \ w\text{-}D^{\alpha-1}u_{\zeta_n}(0) = b. \end{cases}$$

and let u_ν be the unique solution to

$$\begin{cases} w\text{-}D^\alpha u_\nu(t) = \int_Z f\left(t, u_\nu(t), w\text{-}D^{\alpha-1}u_\nu(t), z\right) \nu_t(dz) \\ u_\nu(0) = 0,\ w\text{-}D^{\alpha-1}u_\nu(0) = b. \end{cases}$$

In view of Step 1 of the proof of Theorem 6.2, the sequence (u_{ζ_n}) pointwise converges to u_ν, $(D^{\alpha-1}u_{\zeta_n})$ pointwise converges to $D^{\alpha-1}u_\nu$ so that, $\delta_{u_{\zeta_n}} \otimes \delta_{w\text{-}D^{\alpha-1}u_{\zeta_n}} \otimes \nu^n$ stably converges to $\delta_{u_\nu} \otimes \delta_{w\text{-}D^{\alpha-1}u_\nu} \otimes \nu$. So we have

$$\lim_{n\to\infty} \int_0^1 L\left(t, u_{\zeta_n}(t), w\text{-}D^{\alpha-1}u_{\zeta_n}(t), \zeta_n(t)\right) dt$$
$$= \int_0^1 [\int_Z L\left(t, u_\nu(t), w\text{-}D^{\alpha-1}u_\nu(t), z\right) \nu_t(dz)]dt.$$

As

$$\int_0^1 L\left(t, u_{\zeta_n}(t), w\text{-}D^{\alpha-1}u_{\zeta_n}(t), \zeta_n(t)\right) dt \geq \inf_{\zeta\in\mathcal{Z}} \int_0^1 L(t, u_\zeta(t), w\text{-}D^{\alpha-1}u_\zeta(t), \zeta(t))dt$$

for all $n \in \mathbf{N}$, it follows that

$$\int_0^1 [\int_Z L(t, u_\nu(t), w\text{-}D^{\alpha-1}u_\nu(t), z)d\nu_t(z)]dt$$
$$\geq \inf_{\zeta\in\mathcal{Z}} \int_0^1 L(t, u_\zeta(t), w\text{-}D^{\alpha-1}u_\zeta(t), \zeta(t))dt.$$

As $\nu \in \mathcal{R}$ is arbitrary given, we get

$$\inf_{\nu\in\mathcal{R}} \int_0^1 [\int_Z L(t, u_\nu(t), w\text{-}D^{\alpha-1}u_\nu(t), z)d\nu_t(z)]dt$$
$$\geq \inf_{\zeta\in\mathcal{Z}} \int_0^1 L(t, u_\zeta(t), w\text{-}D^{\alpha-1}u_\zeta(t), \zeta(t))dt.$$

But

$$\inf_{\zeta\in\mathcal{Z}} \int_0^1 L(t, u_\zeta(t), w\text{-}D^{\alpha-1}u_\zeta(t), \zeta(t))dt$$
$$\geq \inf_{\nu\in\mathcal{R}} \int_0^1 [\int_Z L(t, u_\nu(t), w\text{-}D^{\alpha-1}u_\nu(t), z)d\nu_t(z)]dt$$

thus finishing the proof. ■

Comments. The separability of Banach space E is crucial for our purpose and so is the weak compactness assumption of the second member of the (FDI) under consideration. Further related variants can be found in [21] dealing with convex compact valued second member F. Our tools allow to treat many results in fractional differential inclusions with new applications and perspectives.

References

1. Agarwal RP, Arshad S, O' Reagan D, Lupulescu V (2012) Fuzzy fractional integral equations under compactness type condition. Frac Calc App Anal 15(4):572–590
2. Agarwal RP, Ahmad B, Alsaedi A, Shahzad N (2013) Dimension of the solution set for fractional differential inclusion. J Nonlinear Convex Anal 14(2):314–323
3. Ahmad B, Nieto J (2012) Riemann-Liouville fractional differential equations with fractional boundary conditions. Fixed Point Theory 13(2):329–336
4. Amrani A, Castaing C, Valadier M (1998) Convergence in Pettis norm under extreme point condition. Vietnam J Math 26(4):323–335
5. Angosto C, Cascales B (2009) Measures of weak noncompactness in Banach spaces. Topol Appl 156:1412–1421
6. Attouch H, Cabot A, Redont P (2002) The dynamics of elastic shocks via epigraphical regularization of a differential inclusion. Barrier and penalty approximations. Adv Math Sci Appl 12(1):273–306. Gakkotosho, Tokyo
7. Azzam DL, Castaing C, Thibault L (2001) Three boundary value problems for second order differential inclusions in Banach spaces. Control Cybern 31:659–693
8. Balder EJ (2000) Lectures on Young measure theory and its applications in economics. Rend Istit Mat Univ Trieste 31:1–69. Workshop di Teoria della Misura et Analisi Reale Grado, 1997 (Italia)
9. Benchohra M, Henderson J, Ntouyas SK, Ouahab A (2008) Existence results for fractional functional differential inclusions with infinite delay and applications to control theory. Frac Cal Appl Anal 11:35–56

10. Benchohra M, Graef J, Mostefai F-Z (2011) Weak solutions for boundary-value problems with nonlinear fractional differential inclusions. Nonlinear Dyn Syst Theory 11(3):227–237
11. Castaing C (1972) Quelques résultats de compacité liés a l'intégration. CR Acd Sc Paris 270:1732–1735 (1970) et Bulletin Soc Math France 31:73–81
12. Castaing C (1980) Topologie de la convergence uniforme sur les parties uniformément intégrables de L^1_E et théorèmes de compacité faible dans certains espaces du type Köthe-Orlicz. Sém Anal Convexe 10:5.1–5.27
13. Castaing C (1996) Weak compactness and convergences in Bochner and Pettis integration. Vietnam J Math 24(3):241–286
14. Castaing C, Godet-Thobie C (2014) Le Xuan Truong and Satco B, Optimal control problems governed by a second order ordinary differential equation with m-point boundary condition. Adv Math Econ 18:1–56
15. Castaing C, Truong LX (2012) Some topological properties of solutions set in a second order inclusion with m-point boundary condition. Set-Valued Var Anal 20:249–277
16. Castaing C, Truong LX (2013) Bolza, relaxation and viscosity problems governed by a second order differential equation. J Nonlinear Convex Anal 14(2):451–482
17. Castaing C, Valadier M (1977) Convex analysis and measurable multifunctions. Lecture notes in mathematics, vol 580. Springer, Berlin/Heidelberg/New York
18. Castaing C, Raynaud de Fitte P, Valadier M (2004) Young measures on topological spaces. With applications in control theory and probability theory. Kluwer Academic, Dordrecht
19. Castaing C, Ibrahim AG, Yarou M (2008) Existence problems in second order evolution inclusion: discretization and variational approach. Taiwan J Math 12(6):1435–1477
20. Castaing C, Godet-Thobie C, Hoang TD, Raynaud de Fitte P (2015) On the integration of fuzzy level sets. Adv Math Econ 19:1–32
21. Castaing C, Phung PD, Truong LX (Submitted) On a fractional differential inclusion with integral boundary condition in Banach spaces
22. Cernea A (2010) On a fractional differential inclusion with boundary condition. Studia Universitatis Babes-Bolyai Mathematica LV:105–113
23. Cichon M, Salem HAH (2014) Second order Three-point boundary value problems in abstract spaces. Acta Mathematicae Applicatae Sinica 30(4):1131–1152
24. De Blasi FS (1977) On a property of the unit sphere in a Banach space. Bull Math de la Soc Sci Math de la R.S de la Roumanie. Tome 21(64):3–4
25. El-Sayed AMA (1998) Nonlinear functional differential equations of arbitrary orders. Nonlinear Anal 33:181–186
26. El-Sayed AMA, Abd El-Salam ShA (2009) Nonlocal boundary value problem of a fractional-order functional differential equation. Int J Nonlinear Sci 7:436–442
27. El-Sayed AMA, Ibrahim AG (2001) Set-valued integral equations of arbitrary (fractional) order. Appl Math Comput 118:113–121
28. Florescu LC, Godet-Thobie C (2012) Young measures and compactness in measure spaces. De Gruyter, Berlin
29. Grothendieck A (1964) Espaces vectoriels topologiques, 3rd edn Publicacão da Sociedade de Matemática de Sao Paulo, IX, 320p
30. Ibrahim AG, El-Sayed MA (1997) Definite integral of fractional order for Set-Valued function. Fractional Calc Appl Anal 11(3):81–87
31. Kilbas AA, Marichev OI, Samko SG (1993) Fractional integrals and derivatives: theory and applications. Gorden and Breach, New York
32. Laksmikantham F, Mohapatra RN (2003) Theory of fuzzy differential equations. Taylor and Francis, London/New York
33. Miller KS, Ross B (1993) An introduction to the fractional calculus and fractional differential equations. Wiley, NewYork
34. O'Regan D (2000) Fixed point theorem for weakly sequentially closed maps. Archivum Mathematicum (BRNO) Tomus 36:61–70

35. Ouahab A (2008) Some results for fractional boundary value problem of differential inclusions. Nonlinear Anal 69:3877–3896
36. Phung PH, Truong LX (2013) On a fractional differential inclusion with integral boundary conditions in Banach space. Fractional Calc Appl Anal 16(3):538–558
37. Podlubny I (1999) Fractional differential equation. Academic, New York

Adv. Math. Econ. 20, 77–87 (2016)

Advances in
MATHEMATICAL
ECONOMICS

On First-Order Partial Differential Equations: An Existence Theorem and Its Applications

Yuhki Hosoya

Abstract In this paper, we present an equivalence theorem between the existence of a global solution of a standard first-order partial differential equation and the extendability of the solution of corresponding ordinary differential equation. Moreover, we use this result to produce existence theorems on partial differential equation, and apply this theorem to the integrability problem in consumer theory.

Keywords Partial differential equation • Global solution • Nikliborc's theorem • Shephard's lemma • Expenditure function • Integrability

Article Type: Research Article
Received: July 8, 2015
Revised: October 21, 2015

JEL Classification Numbers: D11
MSC Codes: 35A01, 91B08, 91B16, 91B42

Y. Hosoya (✉)
Department of Economics, Kanto-Gakuin University, 1-50-1601, Miyamachi, Fuchu, Tokyo, 183-0023, Japan
e-mail: hosoya@kanto-gakuin.ac.jp

S. Kusuoka, T. Maruyama (eds.), *Advances in Mathematical Economics Volume 20*, Advances in Mathematical Economics,
DOI 10.1007/978-981-10-0476-6_3

1 Introduction

Consider the following partial differential equation:

$$Du(p) = f(p, u(p)).$$

In this paper, we present an equivalence theorem between the existence of a global solution of the above equation and the corresponding **extension** problem for a solution of some ordinary differential equation. Using this result, we can easily reproduce results of [2, 3]. Further, we present an additional existence theorem on a global solution of the above equation, and see that this theorem can be used in the integrability problem in consumer theory.

In Sect. 2.1, we confirm basic knowledges on ordinary differential equations used in this paper. In Sect. 2.2, we present our basic equivalence theorem. In Sect. 2.3, we argue several existence theorems for the above equation. In Sect. 2.4, we discuss the application of this theorem for economics.

2 Main Results

Throughout this paper, we use the following notations. Let $\mathbb{R}^n$ be n-dimensional Euclidean space, and let $x \in \mathbb{R}^n$. Then, x_i denotes i-th coordinate of x. Meanwhile, let $f : P \to \mathbb{R}^n$ be given. If $x = f(p)$, then x_i is also denoted by $f^i(p)$. Next, for any $x, y \in \mathbb{R}^n$, $x \geq y$ means $x_i \geq y_i$ for all i, and $x \gg y$ means $x_i > y_i$ for all i. $\mathbb{R}^n_{++}$ denotes the set of all $x \in \mathbb{R}^n$ such that $x \gg 0$, and called the **positive orthant**. Meanwhile, $\mathbb{R}^n_+$ denotes the set of all $x \in \mathbb{R}^n$ such that $x \geq 0$, and is called the **nonnegative orthant**.

2.1 Preliminaries: Ordinary Differential Equations

In this subsection, we mention several basic results on the solutions of ordinary differential equations. All results in this subsection are verified in many famous textbooks on differential equations.[1]

[1]We mainly refer to Chapter 4 of [4], which is a very good textbook. However, only the 1st edition of this book has been translated into English, and this edition is rather complicated and difficult to understand. In the 2nd edition, this problem is removed, although this has been translated into Japanese [5] but not English. The essence of results in this section are also included in [1, 6], and many other textbooks.

First, consider the following differential equation:

$$\dot{x} = f(t,x), \ x(t_0) = x_0,$$

where $\dot{x}$ means the derivative of x with respect to t. A **solution** of this equation is a C^1 function x defined on an interval I containing t_0 such that $x(t_0) = x_0$ and $\dot{x}(t) = f(t,x(t))$ for any $t \in I$. We assume that f is defined on some open set $P \subset \mathbb{R} \times \mathbb{R}^n$ with $(t_0,x_0) \in P$, and that f is of C^1-class. It is known that there is a solution of the above equation, and for any two solutions x,y, $x(t) = y(t)$ for all t included in the intersection of the domains of x and y. A solution y is called an **extension** of a solution x if and only if the domain of y includes the domain of x. A solution x is **nonextendable** if and only if there is no extension of x except for x itself. Two facts on nonextendable solutions are known. First, there uniquely exists a nonextendable solution defined on an open interval. Second, if x is a nonextendable solution defined on $]a,b[$, then for any compact set $C \subset P$, there exist $\hat{t},\bar{t} \in]a,b[$ such that $(t,x(t)) \notin C$ if either $a < t \le \hat{t}$ or $\bar{t} \le t < b$.

Next, consider the following parametrized differential equation:

$$\dot{x} = f(t,x;y), \ x(t_0;y) = x_0.$$

We assume that f is defined on some open set $\tilde{P} \subset \mathbb{R} \times \mathbb{R}^n \times \mathbb{R}^m$ and that f is of C^1-class with respect to (t,x,y). It is known that there exists a function $x(t;y)$ defined on some open set such that if y is fixed, then $x(t;y)$ is a nonextendable solution of the above problem. We also call this function a **nonextendable solution**. It is also known that x is of C^1-class, and $\frac{\partial^2 x}{\partial t \partial y_i}(t;y)$ can be defined and is equal to $\frac{\partial^2 x}{\partial y_i \partial t}(t;y)$.

2.2 Basic Result

Let $n \ge 2$, P be an open subset of $\mathbb{R}^n \times \mathbb{R}$, Ω be a subset of $\mathbb{R}^n$, $(p,m) \in P$ and $f : P \to \Omega$ be of C^1-class. Define

$$s_{ij}(q,w) = \frac{\partial f^i}{\partial q_j}(q,w) + \frac{\partial f^i}{\partial w}(q,w) f^j(q,w).$$

Then, we say that f is **integrable** if and only if $s_{ij} = s_{ji}$ on P.

Consider the following partial differential equation:

$$Du(q) = f(q,u(q)), \ u(p) = m, \tag{1}$$

and the following ordinary differential equation:

$$\dot{c}(t;p,q) = f((1-t)p + tq, c(t;p,q)) \cdot (q-p), \ c(0) = m. \tag{2}$$

If U is an open set including p, then we say a C^1 function $u : U \to \mathbb{R}$ is a solution of (1) if $u(p) = m$ and $Du(q) = f(q, u(q))$ for any $q \in U$. If U is not open but includes p, then we say a continuous function $u : U \to \mathbb{R}$ is a solution of (1) if there exists an extension $\tilde{u} : \tilde{U} \to \mathbb{R}$ of u in an open set $\tilde{U}$ such that $\tilde{u}$ is a solution of (1).

Theorem 1. *Let $n \geq 2$, P be an open subset of $\mathbb{R}^n \times \mathbb{R}$, Ω be a subset of $\mathbb{R}^n$, $(p, m) \in P$ and $f : P \to \Omega$ be of C^1-class and integrable. Suppose that U is a convex subset of $\mathbb{R}^n$ with $p \in$ int. U. Then, there exists a solution u of (1) defined on U if and only if for any $q \in U$, the solution $c(t; p, q)$ of (2) can be extended on $[0, 1]$. Moreover, in this case such a solution is unique, and $u(q) = c(1; p, q)$ for any $q \in U$.*

Proof. Suppose that $u : U \to \mathbb{R}^n$ is a solution of (1) defined on U. Choose any $q \in U$ and define

$$c(t; p, q) = u((1 - t)p + tq).$$

Then, $c(t; p, q)$ is a solution of (2) defined on $[0, 1]$. In this case, the solution of (2) is unique, and thus u is also unique.

Conversely, suppose that for any $q \in U$, the solution $c(t; p, q)$ can be extended on $[0, 1]$, and define

$$u(q) = c(1; p, q).$$

We need to verify that u is a solution of (1). First, because the domain of the function $(t, q) \mapsto c(t; p, q)$ is open, we can extend u to some open set $\tilde{U}$ including U. Therefore, we assume that U is not necessarily convex but open. It suffices to show that above u is a solution of (1) on such U. Let

$$h^i(t; q) = \frac{\partial c}{\partial q_i}(t; p, q) - tf^i((1 - t)p + tq, c(t; p, q)).$$

Then, $h^i(0; q) = 0$ and,[2]

$$\begin{aligned} \dot{h}^i(t; q) &= \frac{\partial^2 c}{\partial t \partial q_i} - f^i - t\left[\sum_{j=1}^n \frac{\partial f^i}{\partial q_j} \times (q_j - p_j) + \frac{\partial f^i}{\partial w}\dot{c}\right] \\ &= f^i + \sum_{j=1}^n \left(t\frac{\partial f^j}{\partial q_i} + \frac{\partial f^j}{\partial w}\frac{\partial c}{\partial q_i}\right) \times (q_j - p_j) \end{aligned}$$

[2]We abbreviate $f((1 - t)p + tq, c(t; p, q))$ to f, $c(t; p, q)$ to c, and so on.

$$-f^i - t\left[\sum_{j=1}^{n}\frac{\partial f^i}{\partial q_j}\times(q_j-p_j)+\frac{\partial f^i}{\partial w}\sum_{j=1}^{n}f^j\times(q_j-p_j)\right]$$

$$=t\sum_{j=1}^{n}\left(\frac{\partial f^j}{\partial q_i}-s_{ij}\right)\times(q_j-p_j)+\sum_{j=1}^{n}\frac{\partial f^j}{\partial w}\frac{\partial c}{\partial q_i}\times(q_j-p_j)$$

$$=[\frac{\partial c}{\partial q_i}-tf^i]\sum_{j=1}^{n}\frac{\partial f^j}{\partial w}(q_j-p_j)$$

$$=h^i(t;q)\sum_{j=1}^{n}\frac{\partial f^j}{\partial w}(q_j-p_j),$$

where the second equality comes from the equation $\frac{\partial^2 c}{\partial t\partial q_i}=\frac{\partial^2 c}{\partial q_i\partial t}$. Thus, $h^i(t;q)$ is a solution of a linear differential equation $\dot{h}^i=a(t)h^i$ with $h^i(0;q)=0$, where

$$a(t)=\sum_{j=1}^{n}\frac{\partial f}{\partial w}((1-t)p+tq,c(t;p,q))\cdot(q-p).$$

By the uniqueness of the solution, we conclude that $h^i(t;q)\equiv 0$. Hence, $\frac{\partial c}{\partial q_i}(1;p,q)=f^i(q,c(1;p,q))$, which implies that

$$Du(q)=f(q,u(q)).$$

This completes the proof. ■

2.3 *Nikliborc's Theorem, Hurwicz-Uzawa's Theorem, and Their Extension*

Nikliborc [3] proved the following result[3]:

Theorem 2 (Nikliborc). *Suppose that* $f:P\to\Omega$ *satisfies all the assumptions in Theorem 1, where* P *includes the cube* $\Pi=(p,m)+\prod_{i=1}^{n}[-a,a]\times[-b,b]$. *Let*

[3]This formulation of the Nikliborc's theorem is in [2]. Actually, Nikliborc's theorem holds under only differentiability and local Lipschitz condition of f. However, our Theorem 1 cannot be applied under such assumptions, because the equation $\frac{\partial^2 c}{\partial t\partial q_i}=\frac{\partial^2 c}{\partial q_i\partial t}$ used in the proof no longer holds.

$Q = p + \prod_{i=1}^n [-c, c]$ and $c \le \min\{a, b/nM\}$, where

$$M = \sup_{i\in\{1,\dots,n\},(q,w)\in\Pi} |f^i(q,w)|.$$

Then, there exists a unique solution u of (1) defined on Q.

In fact, using our result this theorem can be shown easily.

Proof. By Theorem 1, it suffices to show that the solution $c(t; p, q)$ of (2) can be extended to $[0, 1]$ for all $q \in Q$. Suppose not, and let $c(t; p, q)$ can be defined only on $[0, t^*[$, where $t^* \le 1$. Define

$$\bar{t} = \sup\{t \ge 0 | \forall s \in [0, t], |c(s; p, q) - m| \le b\}.$$

Then, $\bar{t} \in [0, t^*]$. Now, for any $t \in [0, \bar{t}]$,

$$\begin{aligned} |c(t; p, q) - m| &= |c(t; p, q) - c(0; p, q)| \\ &\le \int_0^t \left| \sum_{i=1}^n f^i((1-s)p + sq, c(s; p, q))(q_i - p_i) \right| ds \\ &\le \int_0^t \sum_{i=1}^n M|q_i - p_i| ds \\ &\le bt. \end{aligned}$$

If $\bar{t} < t^*$, then $\bar{t} < 1$, and thus $|c(\bar{t}; p, q) - m| < b$. However, in this case $|c(t; p, q) - m| < b$ for any t such that $\bar{t} < t < t^*$, which contradicts the definition of $\bar{t}$. Therefore, $\bar{t} = t^*$. Thus, $(p(t), c(t; p, q)) \in \Pi$ for any $t \in [0, t^*[$, which contradicts the compactness of Π and the definition of t^*. This completes the proof. ■

Meanwhile, Hurwicz and Uzawa [2] modified this theorem to apply it to economics.[4]

Theorem 3 (Hurwicz-Uzawa). *Suppose that $f : P \to \Omega$ is of C^1-class and integrable, where $P = \mathbb{R}^n_{++} \times \mathbb{R}_+$ and $\Omega = \mathbb{R}^n_+$.[5] Moreover, suppose that $f(q, 0) = 0$ for any q, and that for any $q, q' \in \mathbb{R}^n_{++}$ with $q \ll q'$, there exists $M_{q,q'} > 0$ such that $\|\frac{\partial f}{\partial w}(r, w)\| \le M_{q,q'}$ for any $(r, w) \in P$ with $q \le r \le q'$. Then, for any $(p, m) \in P$, there exists a unique solution u of (1) defined on $\mathbb{R}^n_{++}$.*

This result is also easily verified by using our Theorem 1.

---

[4]They claimed that this theorem holds even if f is not C^1 but only differentiable. However, we doubt this claim.

[5]Here, P is not open and thus this assumption means that f can be extended to some open set including P and this extension is of C^1-class and integrable.

Proof. If $m = 0$, then $u(q) \equiv 0$ satisfies (1), and thus we assume that $m > 0$.

First, we define $\tilde{f} : \mathbb{R}^n_{++} \times \mathbb{R} \to \mathbb{R}^n$ by

$$\tilde{f}(q, w) = \begin{cases} f(q, w) & \text{if } w \geq 0, \\ -f(q, -w) & \text{if } w < 0, \end{cases}$$

and consider the following differential equation:

$$Du(q) = \tilde{f}(q, u(q)), \ u(p) = m. \tag{3}$$

We can easily show that $\tilde{f}$ is of C^1-class and integrable. By Theorem 1, the existence of a solution $u : \mathbb{R}^n_{++} \to \mathbb{R}$ of (3) is equivalent to the extendability of the solution $c(t; p, q)$ of the following differential equation:

$$\dot{c}(t; p, q) = \tilde{f}((1 - t)p + tq, c(t; p, q)) \cdot (q - p), \ c(0; p, q) = m,$$

to $[0, 1]$. Suppose that $c(t; p, q)$ is a nonextendable solution of the above equation defined on $]t_*, t^*[$, where $t_* < 0 < t^*$. If $c(t; p, q) = 0$ for some $t \in [0, t^*[$, then by the uniqueness of the nonextendable solution, $c(s; p, q) \equiv 0$, which contradicts the fact that $c(0; p, q) = m > 0$. Therefore, we have $c(t; p, q) > 0$ for all $t \in [0, t^*[$.

Now, suppose that $t^* \leq 1$. Then, $(p(t), c(t; p, q))$ is excluded from any compact set in $\mathbb{R}^n_{++} \times \mathbb{R}$ if $t^* - t$ is sufficiently small, and thus $\limsup_{t \to t^*} c(t; p, q) = +\infty$. Define $p(t) = (1 - t)p + tq$, and choose $q_1, q_2 \in \mathbb{R}^n_{++}$ so that $q_1 \leq p, q \leq q_2$ and $q_1 \ll q_2$. Then,

$$\begin{aligned} |c(t; p, q) - m| &\leq \int_0^t |f(p(s), c(s; p, q)) \cdot (q - p)| ds \\ &\leq \int_0^t \|q - p\| [\|f(p(s), m)\| + \|f(p(s), c(s; p, q)) - f(p(s), m)\|] ds \\ &\leq \int_0^t [\|q - p\| M_{q_1, q_2} |c(s; p, q) - m| + \|q - p\| \max_{s' \in [0,1]} \|f(p(s'), m)\|] ds. \end{aligned}$$

We now require the following lemma.

Lemma 1. *Suppose that a continuous function $u : [0, \bar{t}] \to \mathbb{R}$ satisfies*

$$u(0) = 0, \ u(t) \leq \int_0^t [Au(s) + B] ds,$$

for some $A, B > 0$. Then,

$$u(t) \leq \frac{B}{A}(e^{At} - 1).$$

Proof of Lemma 1. Let $u_0(t) = u(t)$, and when $u_k(t)$ is defined, then define

$$u_{k+1}(t) = \int_0^t [Au_k(t) + B]ds.$$

Then, $u_k(t)$ is increasing in k. In fact, this sequence (u_k) is a path of the Picard's iteration on the following linear differential equation:

$$\dot{v}(t) = Av(t) + B, v(0) = 0.$$

Therefore, (u_k) converges uniformly to the solution v of the above differential equation. To solve above equation, we can find $v(t) = \frac{B}{A}(e^{At} - 1)$, and thus $u(t) \le v(t)$. ■

We return to the proof of theorem. By Lemma 1, we have

$$|c(t;p,q) - m| \le \frac{B}{A}(e^{At} - 1),$$

for some $A, B > 0$. This implies that $\limsup_{t \to t^*} c(t;p,q) < +\infty$, a contradiction.

Therefore, there exists a solution $u : \mathbb{R}^n_{++} \to \mathbb{R}$ of (3). Moreover, because $u(q) = c(1;p,q) > 0$, u is also a solution of (1). Obviously, this solution is unique. This completes the proof. ■

However, the existence condition for the Lipschitz-like constant $M_{q,q'}$ in the above theorem is somewhat strange in the context of economics. Therefore, we want to remove this odd requirement. It is natural in economics that $q \cdot f(q,w) = w$ for any $(q,w) \in P$, and this condition is called the **Walras' law**.[6] If this condition holds, then the condition $f(q,0) = 0$ is redundant and we can omit points $(q,0)$ from P. Thus, we can assume that $P = \mathbb{R}^n_{++} \times \mathbb{R}_{++}$. Our question is the following: what is the natural requirement on the existence of a solution of (1), with the aid of the Walras' law? The answer is very simple.

Definition 1. Suppose that $f : P \to \Omega$ satisfies all the conditions in Theorem 1. Define the $(n \times n)$-matrix $S_f(q,w)$ such that the (i,j)-th element of this matrix is $s_{ij}(q,w)$. Then, we say that f satisfies (NSD) if and only if $S_f(q,w)$ is always negative semi-definite for any $(q,w) \in P$.

Later we will argue how (NSD) is natural in economics.

Theorem 4. *Suppose that $f : P \to \Omega$ satisfies all the conditions in Theorem 1, where $P = \mathbb{R}^n_{++} \times \mathbb{R}_{++}$ and $\Omega = \mathbb{R}^n_+$. If f satisfies Walras' law and (NSD), then for any $(p,m) \in P$, there uniquely exists a solution $u : \mathbb{R}^n_{++} \to \mathbb{R}_{++}$ of (1).*

[6]Hurwicz and Uzawa [2] also assumed this condition in their results.

Proof. By Theorem 1, this is equivalent to the extendability of the solution of (2) to $[0, 1]$. Suppose that $c(t; p, q)$ is extendable only on $[0, t^*[$, where $t^* \leq 1$. Then, for any compact set $C \subset \mathbb{R}_{++}$, there exists $\bar{t} \in [0, t^*[$ such that $c(t; p, q) \notin C$ for any $t \in [\bar{t}, t^*[$. This implies that either $\liminf_{t \to t^*} c(t; p, q) = 0$ or $\limsup_{t \to t^*} c(t; p, q) = +\infty$.

Define $p(t) = (1 - t)p + tq$ and $x = f(p, m)$. We need the following lemma:

Lemma 2. *Suppose that $c(t; p, q)$ is a solution of (2) defined on $[0, \hat{t}]$. Let $y = f(p(\hat{t}), c(\hat{t}; p, q))$. Then, $p \cdot y \geq m$ and $p(\hat{t}) \cdot x \geq c(\hat{t}; p, q)$.*

Proof of Lemma 2. Let $d(t) = p \cdot f(p(t), c(t; p, q))$. Then, by simple calculation,

$$\dot{d}(t) = p^T S_f(p(t), c(t; p, q))(q - p).$$

Meanwhile, by Walras' law we can easily shown that

$$p(t)^T S_f(p(t), c(t; p, q))(q - p) = 0.$$

Therefore,

$$\dot{d}(t) = -t(q - p)^T S_f(p(t), c(t; p, q))(q - p) \geq 0.$$

Therefore, we have $p \cdot y = d(\hat{t}) \geq d(0) = m$. The proof of the rest claim is symmetrical and we omit it. ■

By Lemma 2, we have $p(t) \cdot x \geq c(t; p, q)$, and thus $\limsup_{t \to t^*} c(t; p, q) < +\infty$. Therefore, $\liminf_{t \to t^*} c(t; p, q) = 0$, and thus there exists a sequence (t^k) such that $t^k \uparrow t^*$ and $c(t^k; p, q) \to 0$ as $k \to \infty$. Let $x^k = f(p(t^k), c(t^k; p, q))$. By Lemma 2, we have $p \cdot x^k \geq m = p \cdot x$ and $p(t^k) \cdot x \geq c(t^k; p, q) = p(t^k) \cdot x^k$. This implies that $q \cdot x^k \leq q \cdot x$, and thus the sequence (x^k) is bounded. Therefore, taking subsequence, we can assume that $x^k \to x^* \in \mathbb{R}^n_+$. Because $p \cdot x^* \geq m$, we have $x^* \neq 0$. Then,

$$0 < p(t^*) \cdot x^* = \lim_{k \to \infty} p(t^k) \cdot x^k = \lim_{k \to \infty} c(t^k; p, q) = 0,$$

a contradiction. This completes the proof. ■

2.4 Application in Economics: The Integrability Problem

Consider the following (simple) optimization problem:

$$\begin{aligned} \max \quad & v(x) \\ \text{subject to.} \quad & x \geq 0, \\ & p \cdot x \leq m, \end{aligned}$$

where x is called the **consumption bundle**, p is called the **price vector**, and m is called the **money income**. The value $v(x)$ measures the goodness of the consumption bundle x for this consumer, and the function v is called the **utility function**. This problem represents the usual consumption problem. Under several conditions, for any $p \gg 0$ and $m > 0$ there uniquely exists a solution $f(p,m)$ of the above problem, and the function f is called the **demand function** of v. If v is increasing, then clearly f satisfies Walras' law.

Now, suppose that f is the C^1 demand function of v. Choose any $(p,m) \in \mathbb{R}^n_{++} \times \mathbb{R}_{++}$ and let $x = f(p,m)$. Consider the following function:

$$E^x(q) = \inf\{q \cdot y | v(y) \ge v(x)\}.$$

The function E^x is called the **expenditure function**. If v is monotone, then we can easily verify that $E^x(p) = m$ and E^x is a concave function. Further, the following result is known, and called Shephard's lemma[7]:

$$DE^x(q) = f(q, E^x(q)).$$

Therefore, the function E^x is a solution of (1). Note that $S_f(p,m) = D^2E^x(p)$. Therefore, we have f satisfies both the integrability condition and (NSD).

Conversely, suppose that f is a C^1 function from $\mathbb{R}^n_{++} \times \mathbb{R}_{++}$ into $\mathbb{R}^n_+$ that satisfies the integrability condition, (NSD), and Walras' law. Can we ensure the existence of the utility function v such that f is the demand function of v? The answer is **yes**.

The key idea is the following. Let f be some demand function of v. Let x, y be included in the range of f, and define E^x, E^y as above. Then, for any $\bar{p} \in \mathbb{R}^n_{++}$, we have $E^x(\bar{p}) \ge E^y(\bar{p})$ if and only if $v(x) \ge v(y)$. Therefore, if we fix $\bar{p}$ and define

$$w(x) = \begin{cases} E^x(\bar{p}) & \text{if } x \text{ is in the range of } f, \\ 0 & \text{otherwise,} \end{cases}$$

then it is natural that f is the demand function of w.

This is the case in which v is known. If v is unknown, how do we define E^x? To solve this, we can use the Shephard's lemma. That is, we can define E^x as the unique solution of (1), where (p,m) satisfies that $x = f(p,m)$. Then, we can define w as above. Hurwicz and Uzawa [2] showed that if f satisfies Walras' law and (NSD), and there exists a global solution $u : \mathbb{R}^n_{++} \to \mathbb{R}_{++}$ of (1) for any initial condition, then f is the demand function of w, where

$$w(x) = \begin{cases} u(\bar{p}) & \text{if } x = f(p,m) \text{ and } u \text{ is a solution of (1) with } u(p) = m, \\ 0 & \text{if } x \neq f(p,m) \text{ for any } (p,m). \end{cases}$$

[7] See Theorem 1 of [2].

However, they could not verify the existence of the solution u of (1) under only Walras' law, (NSD) and integrability condition. Hence, they assumed the existence condition for $M_{q,q'}$.

Meanwhile, we have removed this condition from the existence theorem (Theorem 4), and thus we immediately obtain the following result.

Theorem 5. *Suppose that f is C^1, integrable function satisfying Walras' law and (NSD), and choose any $\bar{p} \in \mathbb{R}^n_{++}$ and define w as above. Consider the following problem:*

$$\begin{aligned} \max \quad & w(x) \\ \text{subject to. } & x \geq 0, \\ & p \cdot x \leq m, \end{aligned}$$

where $(p, m) \in \mathbb{R}^n_{++} \times \mathbb{R}_{++}$. Then, $f(p, m)$ is the unique solution of above problem.

Corollary. *Suppose that $f : \mathbb{R}^n_{++} \times \mathbb{R}_{++} \to \mathbb{R}^n_+$ is of C^1-class and satisfies Walras' law. Then, f is a demand function of some utility function if and only if f satisfies (NSD) and the integrability condition.*

References

1. Hartman P (1982) Ordinary differential equations. Birkhäuser, Basel
2. Hurwicz L, Uzawa H (1971) On the integrability of demand functions. In: Chipman JS, Hurwicz L, Richter MK, Sonnenschein HF (eds) Preferences, utility and demand, Harcourt Brace Jovanovich, Inc., New York, pp 114–148
3. Nikliborc W (1929) Sur les équations linéaires aux différentielles totales. Studia Mathematica 1:41–49
4. Pontryagin LS (1962) Ordinary differential equations. Addison-Wesley, Reading (translated from Russian)
5. Pontryagin LS (1968) Ordinary differential equations, 2nd edn. Kyoritsu Shuppan, Tokyo (in Japanese)
6. Smale S, Hirsch MW (1974) Differential equations, dynamical systems, and linear algebra. Academic Press, New York

Adv. Math. Econ. 20, 89–99 (2016)

Advances in
MATHEMATICAL
ECONOMICS

Real Radicals and Finite Convergence of Polynomial Optimization Problems

Yoshiyuki Sekiguchi

Abstract Polynomial optimization appears various areas of mathematics. Although it is a fully nonlinear nonconvex optimization problems, there are numerical algorithms to approximate the global optimal value by generating sequences of semidefinite programming relaxations. In this paper, we study how real radicals of ideals have roles in duality theory and finite convergence property. Especially, duality theory is considered in the case that the truncated quadratic module is not necessarily closed. We will also try to explain the results by giving concrete examples.

Keywords Polynomial optimization • Real radicals • Sums of squares • Moment problems

Article Type: Research Article
Received: July 31, 2015
Revised: November 2, 2015

JEL Classification: C61
Mathematics Subject Classification (2010): 90C46, 13J30

Y. Sekiguchi (✉)
Graduate School of Marine Science and Technology, Tokyo University of Marine Science and Technology, 2-1-6 Etchujima, Koto, Tokyo, Japan
e-mail: yoshi-s@kaiyodai.ac.jp

S. Kusuoka, T. Maruyama (eds.), *Advances in Mathematical Economics Volume 20*, Advances in Mathematical Economics,
DOI 10.1007/978-981-10-0476-6_4

1 Introduction

Let $\mathbb{R}[x]$ be the ring of polynomials with real coefficients in n variables $x = (x_1, \ldots, x_n)$. For given $f, g_i, h_j \in \mathbb{R}[x]$,

$$\begin{aligned} \text{(POP)} \quad & \inf \quad f(x) \\ & \text{s.t.} \quad g_i(x) \geq 0, \; i = l, \ldots, l, \\ & \qquad\; h_j(x) = 0, \; j = 1, \ldots, m \end{aligned}$$

is called *polynomial optimization problems*. The feasible set of (POP)

$$S := \{x \in \mathbb{R}^n \mid g_i(x) \geq 0, \; h_j(x) = 0, \; i = 1, \ldots, l, j = 1, \ldots, m\}$$

is called a *semialgebraic set*. The polynomial optimization problems appears various areas of mathematics, engineering and economics; see e.g. [1, 3, 4].

Although (POP) is a fully nonlinear nonconvex optimization problem, there are numerical algorithms to approximate the *global* optimal value by generating sequences of semidefinite programming relaxations. The core of the algorithms is the properties obtained from a purely algebraic question; whether given nonnegative polynomials can be written as sums of square polynomials. To test nonnegativity of polynomials is a hard problem but to test whether a polynomial is a sum of square polynomials is a problem solved by semidefinite programming. In addition, representation theorems for nonnegative polynomials from real algebraic geometry ensures the convergence of the algorithms to the global optimal value to arbitrary precision.

From the view of optimization theory, there are two basic problems. One is duality theory for generated semidefinite programs and the other is the *finite convergence property*. Theoretically, the algorithm needs to solve infinitely many semidefinite programs. However it often happens that finite relaxations are enough to obtain the global optimal value (finite convergence property). This phenomenon is known to be closely related to optimality conditions and constraint qualifications [6].

In this paper, we study how real radicals of ideals have roles in duality theory and the finite convergence property. Especially, duality theory is considered in the case that the truncated quadratic module is not necessarily closed. We will also try to explain the results by giving concrete examples.

2 Preliminaries

Semidefinite Programming

Let $\mathbb{S}^n$ be the set of $n \times n$ symmetric matrices and $\mathbb{S}^n_+$ be the set of $n \times n$ positive semidefinite matrices. For $A = (a_{ij}), B = (b_{ij}) \in \mathbb{S}^n$, the *inner product* is defined by $\langle A, B \rangle = \sum_{i,j=1}^n a_{ij} b_{ij}$. For given $A_i, C \in \mathbb{S}^n, b_i \in \mathbb{R}$, the problems with the form of

$$(\mathscr{P}) \quad \sup \quad -\langle C, X \rangle \quad \text{s.t.} \quad \langle A_i, X \rangle = b_i, \quad i = 1, \ldots, m, \quad X \in \mathbb{S}^n_+$$

is called a *semidefinite programming problem*. The dual problem is

$$(\mathscr{D}) \quad \inf \quad b^T y \quad \text{s.t.} \quad C + \sum_{i=1}^m A_i y_i \in \mathbb{S}^n_+.$$

We denote the optimal values of $(\mathscr{P})$ and $(\mathscr{D})$ by $\sup(\mathscr{P})$ and $\inf(\mathscr{D})$ and so on. Then the following duality theorem holds.

Theorem 2.1 ([11]). *Suppose the feasible sets of* $(\mathscr{P})$ *and* $(\mathscr{D})$ *are nonempty. If the feasible set of* $(\mathscr{P})$ *has an interior point,* $\sup(\mathscr{P}) = \inf(\mathscr{D})$ *and* $(\mathscr{D})$ *has an optimal solution.*

Sums of Square Polynomials

For $h_i \in \mathbb{R}[x], f = \sum_{i=1}^m h_i^2$ is called a *sum of square polynomials* (SOS). Let $\sum \mathbb{R}[x]^2$ be the set of sums of square polynomials. Then SOS polynomial f is nonnegative, but nonnegative polynomial is not necessarily SOS; see e.g. [5].

For a given $f \in \mathbb{R}[x]$, consider

$$(\mathscr{P}) \quad \inf \quad f(x), \ x \in \mathbb{R}^n.$$

It is equivalent to

$$\sup \quad r \quad \text{s.t.} \quad f(x) - r \geq 0, \ \forall x \in \mathbb{R}^n.$$

Then this problem has infinitely many constraints (semi-infinite programming) and hard to solve. However the following relaxation problem is easier to solve:

$$\inf(\mathscr{P}) \quad \geq \quad \sup \quad r \quad \text{s.t.} \quad f(x) - r \in \sum \mathbb{R}[x]^2.$$

To explain it, we introduce the following notation. For $\alpha = (\alpha_1, \ldots, \alpha_n) \in \mathbb{N}^n$, let $x^\alpha = x_1^{\alpha_1} \cdots x_n^{\alpha_n}$ and $|\alpha| = \sum_{i=1}^n \alpha_i$. The *degree* $\deg f$ of f is the maximum of the degrees of monomials in f.

Proposition 2.2. *Let $f \in \mathbb{R}[x]$ with $\deg f = 2d$. f is a sum of square polynomials if and only if there exists positive semidefinite matrix X such that $f = u^T X u$, where $u = (x^\alpha)_{|\alpha| \leq d}$.*

Real Algebra

For $h_1, \ldots, h_m \in \mathbb{R}[x]$, let $h = (h_1, \ldots, h_m)$. Then the *ideal* $\langle h \rangle$ generated by $h_1, \ldots, h_m$ is $\langle h \rangle = \{a_1 h_1 + \cdots a_m h_m \mid a_j \in \mathbb{R}[x]\}$. For $k \in \mathbb{N}$,

$$\langle h \rangle_{2k} = \{a_1 h_1 + \cdots a_m h_m \mid a_j \in \mathbb{R}[x], \deg(a_j h_j) \leq 2k\}$$

is called the *2k–truncated ideal*. The *variety* $\mathscr{V}_{\mathbb{R}}(I)$ of I in $\mathbb{R}^n$ is $\mathscr{V}_{\mathbb{R}}(I) = \{x \in \mathbb{R}^n \mid p(x) = 0, \ \forall p \in I\}$. For a variety $V \subset \mathbb{R}^n$, the *vanishing ideal* $\mathscr{I}(V)$ is $\mathscr{I}(V) = \{p \in \mathbb{R}[x] \mid p(x) = 0, \ \forall x \in V\}$. The ideal $I \in \mathbb{R}[x]$ is said to be *real* if $\mathscr{I}(\mathscr{V}_{\mathbb{R}}(I)) = I$. In general, $\mathscr{I}(\mathscr{V}_{\mathbb{R}}(I))$ coincides with the *real radical* $\sqrt[\mathbb{R}]{I}$ of I given by

$$\sqrt[\mathbb{R}]{I} = \left\{p \in \mathbb{R}[x] \mid p^{2\ell} + \sigma \in I \text{ for some } \ell \in \mathbb{N}, \ \sigma \in \sum \mathbb{R}[x]^2\right\}.$$

For $g_1, \ldots, g_l \in \mathbb{R}[x]$ and $k \in \mathbb{N}$,

$$M(g) = \left\{\sigma_0 + \sum_i \sigma_i g_i \mid \sigma_i \in \sum \mathbb{R}[x]^2\right\},$$

is called a *quadratic module* and

$$M_k(g) = \left\{\sigma_0 + \sum_j \sigma_j g_j \mid \sigma_j \in \sum \mathbb{R}[x]^2, \deg \sigma_0, \deg(\sigma_j g_j) \leq 2k\right\},$$

is called the *k-th truncated quadratic module*. We also call $M_k + I_{2k}$ k-th truncated quadratic module.

Lasserre's Hierarchy

Consider

$$\text{(POP)} \quad \inf \quad f(x) \quad \text{s.t.} \quad g_i(x) \geq 0,\ h_j(x) = 0.$$

Let $S = \{x \in \mathbb{R}^n \mid g_i(x) \geq 0,\ h_j(x) = 0\}$. Then obviously we have

$$(*) \quad q \in M(g) + \langle h \rangle \Rightarrow q \geq 0 \text{ on } S$$

Thus

$$\begin{aligned} \inf(\text{POP}) &= \inf \quad f(x) \quad \text{s.t.} \quad h_i(x) = 0,\ g_j(x) \geq 0 \\ &= \sup \quad r \quad \text{s.t.} \quad f(x) - r \geq 0 \text{ on } S \\ &\geq \sup \quad r \quad \text{s.t.} \quad f(x) - r \in M(g) + \langle h \rangle \\ \geq \sup(\text{SOS}_k) &:= \sup \quad r \quad \text{s.t.} \quad f(x) - r \in M_k(g) + \langle h \rangle_{2k}. \end{aligned}$$

The last problem can be written by a semidefinite program and is called a *k-th SOS relaxation* for (POP). Here it is natural to ask that for $q \in \mathbb{R}[x]$, whether $q \geq 0$ on S implies that $q \in M(g) + \langle h \rangle$.

For a quadratic module M, M is said to be *Archimedean*, if $\forall p \in \mathbb{R}[x],\ \exists N \in \mathbb{N}$ such that $N \pm p \in M$. In fact, if $N - \sum_{i=1}^n x_i \in M$, then M is Archimedean. So, for a compact semialgebraic set S, the associated quadratic module can be modified to an Archimedean quadratic module by adding $N - \sum_{i=1}^n x_i$ without changing S. The following is a representation theorem for positive polynomials.

Theorem 2.3 ([8]). *Suppose that $M(g) + \langle h \rangle$ is Archimedean. If $p \in \mathbb{R}[x]$ satisfies $p > 0$ on S, then $p \in M(g) + \langle h \rangle$.*

Thus the first inequality is actually an equality and then

Theorem 2.4 ([2]). *If $M(g) + \langle h \rangle$ is Archimedean, then* $\sup(\text{SOS}_k) \to \inf(\text{POP})$ *as $k \to \infty$.*

The dual problem of (SOS_k) is called the *moment problem*

$$\begin{aligned} (\text{MOM}_k) \quad \inf \quad & L(f) \\ \text{s.t.} \quad & L : \mathbb{R}[x] \to \mathbb{R},\ \text{linear},\ L(1) = 1, \\ & L(p) \geq 0, \forall p \in M_k(g) + \langle h \rangle_{2k} \end{aligned}$$

and this is also written by a semidefinite program. In addition, we have the weak duality theorem

$$\inf(\text{POP}) \geq \inf(\text{MOM}_k) \geq \sup(\text{SOS}_k).$$

For the details, see [3–5, 9].

3 Finite Convergence Property

In the rest of the paper, we assume that

$$\text{(POP)} \quad \inf \quad f(x) \quad \text{s.t.} \quad g_i(x) \geq 0,\ h_j(x) = 0.$$

has optimal value and inf(POP) $= 0$.

In theory, to obtain inf(POP), we need to solve infinitely many semidefinite programs (SOS_k). However, in practice, it often happens that $\sup(\text{SOS}_{k_0}) = \inf(\text{POP})$ for some $k_0 \in \mathbb{N}$. In this event, (POP) is said to have the *finite convergence property*. Clearly, (POP) has the finite convergence property, if $f \in M(g) + \langle h \rangle$. However there is a (POP) which has the finite convergence property but $f \notin M(g) + \langle h \rangle$. A sufficient condition is given in the following:

Theorem 3.1. *Suppose* $\mathscr{I}(S) = \mathscr{I}(\mathscr{V}_{\mathbb{R}}(\langle h \rangle))$. *If* $f \in M(g) + \sqrt[\mathrm{R}]{\langle h \rangle}$, *Problem* POP *has the finite convergence property.*

Proof. This is a reformulation of [7, Theorem 3.1].

Example 1.

$$\begin{aligned}
\text{(POP)} \quad \inf \quad & f(x,y,z) := x \\
\text{s.t.} \quad & z \geq 0, \\
& h(x,y,z) := (x+2y+z)^2 + (x+y)^2 + (y+z)^2 = 0
\end{aligned}$$

Here the feasible region is $\{(t,-t,t) \mid t \geq 0\}$ and the optimal value is $f(0,0,0) = 0$. Let $M := M(z) = \{\sigma_0 + \sigma_1 z \mid \sigma_i \in \sum \mathbb{R}[x,y,z]^2\}$. Then $f \notin M + \langle h \rangle$ but $f \in M + \sqrt[\mathrm{R}]{\langle h \rangle}$. In fact,

$$\sqrt[\mathrm{R}]{\langle h \rangle} = \mathscr{I}(\mathscr{V}_{\mathbb{R}}(\langle h \rangle)) = \mathscr{I}(\{(t,-t,t) \mid t \in \mathbb{R}\}) = \langle x-z, y+z \rangle$$

and

$$x = 1^2 \cdot z + (x-z) \in M + \sqrt[\mathrm{R}]{\langle h \rangle}.$$

On the other hand, suppose that

$$x = \sigma_0 + \sigma_1 z + a\left\{(x+2y+z)^2 + (x+y)^2 + (y+z)^2\right\}$$

for some $\sigma_0, \sigma_1 \in \sum \mathbb{R}[x]^2$, $a \in \mathbb{R}[x]$. Since h is a homogeneous polynomial with degree 2 and $\sigma_1 z$ has a divisor z, we have that only σ_0 can produce x. In addition to produce x from a sum of square polynomials, σ_0 has to have a constant term. However the constant term can not be canceled by other terms. Therefore f can not be represented by an elements of $M+\langle h\rangle$, but the problem has the finite convergence property.

Now the SOS relaxation with degree 2 is the following:

$$
\begin{aligned}
(\mathrm{SOS}_1)\ \sup\quad & r \\
\text{s.t.}\quad & x - r = \begin{pmatrix}1\ x\ y\ z\end{pmatrix} X_0 \begin{pmatrix}1\\ x\\ y\\ z\end{pmatrix} + X_1 z + X_2 h(x,y,z), \\
& X_0 \in \mathbb{S}^4_+,\ X_1 \geq 0,\ X_2 \in \mathbb{R}
\end{aligned}
$$

$$
\begin{aligned}
= \sup\quad & r \\
\text{s.t.}\quad & x - r = \langle X_0, \begin{pmatrix}1 & x & y & z\\ x & x^2 & xy & xz\\ y & xy & y^2 & yz\\ z & xz & yz & z^2\end{pmatrix}\rangle + X_1 z \\
& \qquad + X_2\left(2x^2 + 6y^2 + 2z^2 + 6xy + 6yz + 2zx\right), \\
& X_0 \in \mathbb{S}^4_+,\ X_1 \geq 0,\ X_2 \in \mathbb{R}.
\end{aligned}
$$

By comparing coefficients of $x-r$ and the right hand side, we can eliminate X_0 and then we have

$$
\begin{aligned}
= \sup\quad & r \\
\text{s.t.}\quad & \begin{pmatrix}-r & \frac{1}{2} & 0 & -\frac{X_1}{2}\\ \frac{1}{2} & -2X_2 & -3X_2 & -X_2\\ 0 & -3X_2 & -6X_2 & -3X_2\\ -\frac{X_1}{2} & -X_2 & -3X_2 & -2X_2\end{pmatrix} \in \mathbb{S}^4_+,\ X_1 \geq 0,\ X_2 \in \mathbb{R}.
\end{aligned}
$$

We will show that the optimal value of the last problem is 0 and it is not attainable. Let us consider the dual problem:

$$
\begin{aligned}
(\mathrm{MOM}_1)\ \inf\quad & L(f) \\
\text{s.t.}\quad & L(1) = 1,\ L(p) \geq 0,\ \forall p \in M_1 + \langle h\rangle_2, \\
= \inf\quad & \lambda_{100}
\end{aligned}
$$

$$\text{s.t.} \quad \begin{pmatrix} 1 & \lambda_{100} & \lambda_{010} & \lambda_{001} \\ \lambda_{100} & \lambda_{200} & \lambda_{110} & \lambda_{101} \\ \lambda_{010} & \lambda_{110} & \lambda_{020} & \lambda_{011} \\ \lambda_{001} & \lambda_{101} & \lambda_{011} & \lambda_{002} \end{pmatrix} \in \mathbb{S}^4_+, \lambda_{001} \geq 0,$$

$$2\lambda_{200} + 6\lambda_{020} + 2\lambda_{002} + 6\lambda_{110} + 6\lambda_{011} + 2\lambda_{101} = 0.$$

Since $(\lambda_\alpha) = 0$ is a feasible solution, the weak duality implies

$$0 \geq \inf(\text{MOM}_1) \geq \sup(\text{SOS}_1).$$

Then for any feasible (r, X_1, X_2) of (SOS_1), we have $r \leq 0$. By positive semidefiniteness, the $(1,2)$–minor satisfies

$$\begin{vmatrix} -r & 1/2 \\ 1/2 & -2X_2 \end{vmatrix} = 2rX_2 - \frac{1}{4} \geq 0.$$

Thus r can not be 0. In addition, $(r, X_1, X_2) = (-\varepsilon, 1, (-2\varepsilon)^{-1})$ is feasible for (SOS_1) for any $\varepsilon > 0$. In fact, the eigenvalues of

$$X_0(\varepsilon) := \begin{pmatrix} \varepsilon & 1/2 & 0 & -1/2 \\ 1/2 & 1/\varepsilon & 3/(2\varepsilon) & 1/(2\varepsilon) \\ 0 & 3/(2\varepsilon) & 3/\varepsilon & 3/(2\varepsilon) \\ -1/2 & 1/(2\varepsilon) & 3/(2\varepsilon) & 1/\varepsilon \end{pmatrix} \tag{1}$$

are $0, 0, 9/(2\varepsilon), (2\varepsilon^2+1)/(2\varepsilon)$, which are obtained by `Maxima`[1]. Therefore without optimal solutions, $\sup(\text{SOS}_1) = 0 = \inf(\text{POP})$ and this means the problem has the finite convergence property.

4 Closedness of Truncated Quadratic Modules

To solve semidefinite programs, the interior point methods are usually used. The convergence of the methods is ensured if strong duality of the semidefinite program holds. For (POP), there exist criteria for the generated semidefinite programs to have strong duality.

Theorem 4.1 ([5, 10]). *For* (POP), *if* $\mathscr{I}(S) = \langle h \rangle$, *then* $M_k(g) + \langle h \rangle_{2k}$ *is closed for large* k.

Theorem 4.2 ([5, 10]). *If* $M_k(g) + \langle h \rangle_{2k}$ *is closed, then* $\sup(\text{POP}_k) = \inf(\text{MOM}_k)$.

[1] `Maxima` is a free computer algebra system.

There are (POP) where $\sup(\mathrm{SOS}_k) = \inf(\mathrm{MOM}_k)$ but $M_k(g) + \langle h\rangle_{2k}$ is not closed. Actually we give a criterion for such (POP) in the following:

Theorem 4.3. *Suppose* $\mathscr{I}(S) = \mathscr{I}(\mathscr{V}_{\mathbb{R}}(\langle h\rangle))$. *If the objective function* $f \notin M(g) + \langle h\rangle$ *and* $f \in M(g) + \sqrt[\mathrm{R}]{\langle h\rangle}$, *then* $M_k(g) + \langle h\rangle_{2k}$ *is not closed for large* k *but* $\sup(\mathrm{SOS}_k) = \inf(\mathrm{MOM}_k)$ *for large* k.

Proof. Consider

$$\text{(POP)} \quad \inf \quad f(x) \quad \text{s.t.} \quad g_i(x) \geq 0,\ h_j(x) = 0.$$

By Theorem 3.1, we have $0 = \inf(\mathrm{POP}) = \sup(\mathrm{SOS}_k)$ for some $k \in \mathbb{N}$. Then there exists $s_m \to \sup(\mathrm{SOS}_k)$ $(m \to \infty)$ such that

$$f - s_m = \sigma_{0,m} + \sum_i \sigma_{i,m} g_i + \sum_j a_{j,m} h_j$$

for some $\sigma_{i,m} \in \sum \mathbb{R}[x]^2$ and $a_{j,m} \in \mathbb{R}[x]$. Then $f - s_m \in M_k + I_{2k}$ and $f - s_m \to f \notin M_k + I_{2k}$.

Next, we show $\sup(\mathrm{SOS}_k) = \inf(\mathrm{MOM}_k)$. By weak duality theorem [9], $\inf(\mathrm{POP}) \geq \inf(\mathrm{MOM}_k) \geq \sup(\mathrm{SOS}_k)$. Since [7, Theorem 3.1] ensures that $\sup(\mathrm{SOS}_k) = \inf(\mathrm{POP})$ for some k and thus $\sup(\mathrm{SOS}_k) = \inf(\mathrm{MOM}_k)$.

Example 2. We consider the feasible region S in Example 1. Let $M = \{\sigma_0 + \sigma_1 z \mid \sigma_i \in \sum \mathbb{R}[x,y,z]^2\}$. Then $x \in \left(M + \sqrt[\mathrm{R}]{\langle h\rangle}\right) \setminus (M + \langle h\rangle)$. We will show that $M_1 + \langle h\rangle_2$ is not closed by finding a sequence in $M_1 + \langle h\rangle_2$ which converges to x. Since $(r, X_1, X_2) = (-\varepsilon, 1, (-2\varepsilon)^{-1})$ is feasible for (SOS_1)

$$x - (-\varepsilon) = \begin{pmatrix}1 & x & y & z\end{pmatrix} X_0(\varepsilon) \begin{pmatrix}1\\ x\\ y\\ z\end{pmatrix} + 1 \cdot z$$
$$+ (-2\varepsilon)^{-1}\left(2x^2 + 6y^2 + 2z^2 + 6xy + 6yz + 2zx\right),$$

where $X_0(\varepsilon)$ is given in (1). By Cholesky factorization, we have $X_0(\varepsilon) = LL^T$ where

$$L = \begin{pmatrix} 0 & \sqrt{\varepsilon} & 0 & 0 \\ \sqrt{3}/(2\sqrt{\varepsilon}) & 1/(2\sqrt{\varepsilon}) & 0 & 0 \\ \sqrt{3}/\sqrt{\varepsilon} & 0 & 0 & 0 \\ \sqrt{3}/(2\sqrt{\varepsilon}) & -1/(2\sqrt{\varepsilon}) & 0 & 0 \end{pmatrix}.$$

Thus

$$\begin{aligned}
\left(1\ x\ y\ z\right) X_0(\varepsilon) \begin{pmatrix} 1 \\ x \\ y \\ z \end{pmatrix} &= \left(1\ x\ y\ z\right) LL^T \begin{pmatrix} 1 \\ x \\ y \\ z \end{pmatrix} \\
&= \left(\frac{\sqrt{3}}{2\sqrt{\varepsilon}} x + \frac{\sqrt{3}}{\sqrt{\varepsilon}} y + \frac{\sqrt{3}}{2\sqrt{\varepsilon}} z \right)^2 + \left(\sqrt{\varepsilon} + \frac{1}{2\sqrt{\varepsilon}} x - \frac{1}{2\sqrt{\varepsilon}} z \right)^2
\end{aligned}$$

Therefore $x - (-\varepsilon)$ is contained in $M_1 + \langle h \rangle_2$ for any $\varepsilon > 0$ and $x - (-\varepsilon) \to x \notin M_1 + \langle h \rangle_2$ as $\varepsilon \to 0$.

By Theorem 4.3, we characterize the closure of $M_k(g) + \langle h \rangle_{2k}$.

Theorem 4.4. *Suppose* $\mathscr{I}(S) = \mathscr{I}(\mathscr{V}_{\mathbb{R}}(\langle h \rangle))$. *Then*

$$M(g) + \sqrt[\mathrm{R}]{\langle h \rangle} = \bigcup_{k=1}^{\infty} \overline{(M_k(g) + \langle h \rangle_{2k})}$$

Proof. Let $f \in M + \sqrt[\mathrm{R}]{I}$. By the same arguments of the proof of Theorem 4.3, there exist $k \in \mathbb{N}$ and $\{s_m\} \subset \mathbb{R}$ such that $s_m \to 0$ and $f - s_m \in M_k + I_{2k}$. Thus $f \in \overline{(M_k + I_{2k})}$. We show the other inclusion. Now $M_k + I_{2k} \subset M + \sqrt[\mathrm{R}]{I}$. Since $M + \sqrt[\mathrm{R}]{I}$ is closed by Theorem 4.1, the proof is completed.

References

1. Blekherman G, Parrilo P, Thomas R (2013) Semidefinite optimization and convex algebraic geometry. MOS-SIAM series on optimization, vol 13. SIAM, Philadelphia
2. Lasserre JB (2001) Global optimization with polynomials and the problem of moments. SIAM J Optim 11:796–817
3. Lasserre JB (2010) Moments, positive polynomials and their applications. Imperial College Press, London
4. Laurent M (2009) Sums of squares, moments and polynomial optimization, emerging applications of algebraic geometry. IMA volumes in mathematics and its applications, vol 149. Springer, New York, pp 157–270
5. Marshall M (2008) Positive polynomials and sums of squares. Mathematical surveys and monographs, vol 146. American Mathematical Society, Providence
6. Nie J (2014) Optimality conditions and finite convergence of Lasserre's hierarchy. Math Program 146:97–121
7. Nie J (2013) Polynomial optimization with real varieties. SIAM J Optim 23:1634–1646
8. Putinar M (1993) Positive polynomials on compact semi-algebraic sets. Indiana Univ Math J 42:969–984

9. Schweighofer M (2005) Optimization of polynomials on compact semialgebraic sets. SIAM J Optim 15:805–825
10. Sekiguchi Y, Takenawa T, Waki H (2013) Real ideal and the duality of semidefinite programming for polynomial optimization. Jpn J Ind Appl Math 30:321–330
11. Todd MJ (2001) Semidefinite optimization. Acta Numer 10:515–560

Part II
Expository Review

Adv. Math. Econ. 20, 103–128 (2016)

Advances in
MATHEMATICAL
ECONOMICS

On Differentiated and Indivisible Commodities: An Expository Re-framing of Mas-Colell's 1975 Model

M. Ali Khan and Takashi Suzuki

Abstract With a pure exchange economy and its Walrasian equilibrium formalized as a distribution on the space of consumer characteristics, Mas-Colell [50] showed the existence of equilibrium in a pure exchange economy with differentiated and

---

This research is supported by a Grant-in-Aid for Scientific Research (No. 15K03362) from the Ministry of Education, Culture, Sports, Science and Technology, Japan. This work is part of a joint project on "General Equilibrium Theory with a Saturated Space of Consumers" with Nobusumi Sagara of Hosei University; the authors thank him for all his input in this paper, and express their regret that he did not think it enough to accept co-authorship. A preliminary version was prepared when Khan was visiting the *Shanghai University of Finance and Economics (SHUFE),* and he thanks Haomiao Yu and Yongchao Zhang for stimulating conversation and encouragement. This final version has benefited substantially from an anonymous referee's emphasis on the need for replacing the assumption of overriding desirability of "every homogeneous commodity" to a single money-like commodity; and the Editor's insistence that the authors take a stand on reporting the work as a technical note for a narrow specialized readership or as an expository essay for a general audience.

M. Ali Khan (✉)
Department of Economics, The Johns Hopkins University, Baltimore, MD 21218, USA
e-mail: akhan@jhu.edu

T. Suzuki
Department of Economics, Meiji-Gakuin University, 1-2-37 Shirokanedai, Minato-ku, Tokyo 108–8636, Japan
e-mail: takashisuz@jcom.home.ne.jp

S. Kusuoka, T. Maruyama (eds.), *Advances in Mathematical Economics Volume 20*, Advances in Mathematical Economics,
DOI 10.1007/978-981-10-0476-6_5

indivisible commodities. We present a variant of Mas-Colell's theorem; but more than for its own sake, we use it to expose and illustrate recent techniques due to Keisler-Sun [31], as developed in Khan-Rath-Yu-Zhang [38], to translate a result on a large distributionalized economy (LDE) to a large individualized economy (LIE), when the former can be represented by a saturated or super-atomless measure space of consumers, as formalized in Keisler-Sun [31], Carmona-Podczeck [12] and Podczeck [63] respectively. This also leads us to identify, hitherto unnoticed, open problems concerning symmetrization of distributionalized equilibria of economies in their distributionalized formulations. In relating our result to the antecedent literature, we bring into salience the notions of (i)"overriding desirability of the indivisible commodity," as in Hicks [25], Mas-Colell [51] and Yamazaki [83, 84], and of (ii) "bounded marginal rates of substitution," as in Jones [29, 30] and Ostroy-Zame [60]. Our work also relies heavily on the technical notion of Gelfand integration.

Keywords Exchange economy • Differentiated commodities • Indivisible commodities • Large distributionalized economy • Large individualized economy • Cardinality of the set of consumers

Article Type: Expositional Review
Received: August 11, 2015
Revised: December 2, 2015

1 Introduction

After commenting on the canonical formalization of Walrasian general equilibrium theory that hinges on a finite number of perfectly divisible commodities and a finite number of price-taking consumers, as laid out in the standard texts of Debreu (1959) [15], Nikaido (1968) [59], Arrow-Hahn (1971) [2] and McKenzie (2002) [57], Andreu-Mas-Colell writes:

> In contrast, imperfect competition theory starts with a very different perception of the economic realm; commodities are not homogeneous but subject to differentiation and, consequently, traders enjoy a certain degree of monopoly with respect to the commodities they control. Still, the monopoly power of every single trader is limited by the existence of substitutability relations among commodities; it is a common contention of imperfect competition theory that in a large economy with a large number of mutually substitutive commodities and no 'big' trader, every commodity will be substitutable in the market with infinite elasticity and a perfectly competitive outcome will prevail.

JEL Classification: D51
Mathematics Subject Classification (2010): 28A60, 46N10, 60B05, 91B50

Mas-Colell singles out the Robinsonian (1933) [69] and the Chamberlinian (1933–1965) [14] versions of "imperfect" or "monopolistic" competition, and refers to Samuelson's (1969) [72] account for further discussion. He points out that the conception of differentiated commodities is rooted in a rich and long-longstanding antecedent literature, and that "in one form or another it appears whenever quality differentials (or spatial matters) are focused upon; it can be found in Houthakker (1952) [28], Lancaster (1966) [47], or in the more recent literature on hedonic prices [as in] Rosen (1974) [70] and his references." Indeed, one may now round off Mas-Colell's references by including Theil (1952) [79], Gorman (1956) [20], Becker (1965, 1981) [7, 8], and send the reader to Pollak's authoritative (2012) [65] treatment of the modern field of "household economics" to which this literature gave rise.[1]

Unlike the partial equilibrium emphasis of the modern theory of household economics, Mas-Colell's motivation came from Walrasian general equilibrium theory.[2] In particular, he was after an analogue of Aumann's equivalence theorem between core and Walriasian allocations in a setting of differentiated commodities, along with the assurance that he was not showing the equivalence of two empty sets. However, the crucial point of departure for Mas-Colell's work was his observation that "Usually the space of characteristics will be a subset of $\mathbb{R}^n$; for our purposes, however, and with no further complication of the proof, it suffices that it be a metric space." It is this observation that takes him to model an exchange economy on the commodity space $ca(K)$, the space of signed measures on a compact metric space K, but with the limitation that they take only integer values. He interpreted K as the space of commodity characteristics and each measure on K as a commodity vector, and noted that his setting necessarily involves the conception of a differentiated commodity as an indivisible one.

Regarding this simultaneous consideration of heterogeneity and indivisibility in the formalization of a commodity, Mas-Colell wrote:

> The indivisibility assumption on differentiated commodities may appear odd and, indeed, to the extent that our model includes the non-differentiated one, it is a complicating factor.[3] It is, however, of decisive help with the commodity differentiation aspect of the model; it simplifies matters, not only technically, but also conceptually. Since, as a practical matter, any instance where product differentiation becomes interesting involves indivisibilities, the assumption is probably acceptable. We believe that to encompass differentiated commodities which are available in infinitesimal amounts a substantial rethinking of the model is needed.

---

[1] One may also mention here that Kolm (2010; p. 711) [46] writes, "Adding time constraint to budget constraint was done by Jacques Lesourne before rediscovery by Gary Becker, and the resulting cost of time was applied by others to the optimization of public transportation networks."

[2] For this theory, the reader is also referred to a fascinating recent overview in Arrow (2005) [1]; also see [34, 35] in addition to the texts mention in the first sentence of this essay.

[3] Mas-Colell sights here the work of Dierker, Henry and Broome. This work found subsequent extension in [51] and in [45], building on the work of Yamazaki on non-convex consumption sets, and the papers of Khan-Rashid, and Anderson-Khan-Rashid. Since this work is not relevant to the theorem presented here, we do not burden our bibliography with it, and send the reader to [3, 4].

The conceptual reason is an obvious nod to descriptive realism and presumably also to Hicks (1956) [25].[4] The technical reason goes to the heart of the difficulty of Walrasian equilibrium theory with infinite commodities; namely, the norm interior of the positive orthant of the commodity space, $ca(K)$ in the case at hand, is empty.[5] Given his motivation towards a core equivalence theorem, Mas-Colell had of necessity to work with a model of a pure exchange economy[6] with a continuum of consumers,[7] but what is interesting is that he did not adopt the individualized formulation of Aumann (1964, 1966) [5, 6] or the coalitional formulation of Vind (1964) [82], but rather the 1974 distributionalized formulation of a large exchange economy due to Hart and Kohlberg.[8]

Mas-Colell's pioneering work was followed up by Mas-Colell (1977) [51] and Jones (1983, 1984) [29, 30] in papers that can now be read as a domestication of the basic ideas of 1975: the former held on to indivisibilities but eliminated differentiated commodities, while the latter went the other way; both authors being motivated by wanting to relate the model to the standard literature of Walrasian general equilibrium theory. Mas-Colell worked with a finite number n of indivisible commodities, and a single perfectly divisible commodity, and as such with non-convex consumption sets in $\mathbb{R}^{n+1}$. His focus was on the existence and the determinateness of Walrasian equilibria for such an economy. Jones (1983) [29], on the other hand, worked with $ca(K)$ but removed the integer-valued restriction on the non-negative Borel measures by appealing to a formalization of the concept of a "bounded marginal rate of substitution between the differentiated commodities"; also see Jones (1984) [30]. Jones' existence theorem also did away with the

[4]See Chapter V. In Mas-Colell (1977) [51], an explicit connection is made to McKenzie's (1957) [56] paper which follows Hicks. Mas-Colell writes, "In particular, we postulate that some commodities are perfectly divisible (for simplicity, just one) and we argue that this hypothesis, besides being reasonable, is a *sine qua non;* see McKenzie (1956–1957) [56]." For the relation between Hicks and McKenzie, see Khan-Schlee [40] (2015).

[5]As is well-known to workers in the field, this problem does not arise in the spaces ℓ^{∞} and L_{∞}, see Bewley (1972, 1991) [9, 10] and Suzuki (2013a, 2013b) [76, 77].

[6]As for the erasure of the production sector, it has two consequence for our expository re-visitation of Mas-Colell's model. First, we are forced to ignore the rich strand of the literature associated with Hotelling (1929) [27], Hart [21], Dixit-Stiglitz and their followers on optimum product diversity under monopolistic competition. Secondly, we are also led to ignore the rich literature on production externalities and the convergence of Nash equilibria to Walrasian equilibria; on this see the work of Hart, Pascoa and others, and including that of Mas-Colell himself; see [17, 52, 54] for the latter. To be sure, subsequent work extended the results to economies with production; see, for example [13, 19, 24, 61, 62, 75] among others.

[7]Note that it is the equivalence theorem that already necessitates a continuum; the fact that indivisible commodities potentially disturb the upper hemi-continuity of the individual demand correspondence requires additional assumptions that guarantee the "regularizing effect" of the total demand; see the example in [51] (1977) and reproduced in [45] (1981) to motivate their results. We return to this issue in Sect. 3.2 below.

[8]See Hart-Kohlberg (1974) [23] and Hart-Hildenbrand-Kohlberg (1974) [22]; also see Hildenbrand's 1974 [26] text.

assumption of norm-bounded consumption sets. Thus, both authors bracketed the core equivalence result from their work, and concentrated on the positive theory. What bears emphasis is the fact already mentioned that two[9] papers work with the distributionalized formulation of a large exchange economy. This was done to cope with the failure of the analogue of the Lyapunov theorem and of Fatou's lemma for infinite dimensional commodity spaces.[10] Thus, the authors considered an economy and its Walrasian equilibrium as probability measures on the space of consumer characteristics (individual preferences and endowments), and thereby also avoided what can be seen as extraneous convexity assumption on preferences, and in the case of Mas-Colell (1977) [51], on the consumption set itself.

With this antecedent literature serving as its context, we turn to the contribution of this essay. In a nutshell, we revisit Mas-Colell's 1975 [50] model in the light of subsequent work of Mas-Colell (1977) [51], Jones (1983, 1984) [29, 30] and Ostroy-Zame (1994) [60] on the one hand, and that of Keisler-Sun (2009) [31] and Khan-Rath-Yu-Zhang (2015) [38] on the other; also see [37] and [55]. From the first set of papers, we take the idea of "bounded marginal rates of substitution" in our reformulation of the differentiated part of the commodity space to be a free Abelian group generated by the Dirac measures on K, and with notions of aggregation formalized by Gelfand integration. This, coupled with the application of a technical lemma due to Keisler-Sun,[11] results in substantial clarification of the proof technique. From the second set of papers, we take the fruitful idea of a *saturated (super-atomless)* measure space, and show that existence of an equilibrium for a large distributionalized economy (LDE) can be translated into an equilibrium for a large individualized economy (LIE) when the former is assumed to admit a saturated representation, as delineated in Khan et al. (2015) [36], rather than simply an atomless representation, as in Mas-Colell (1975) [50].

The plan of the essay then is straightforward. Section 2 presents the model and the existence result for the LDE; Sect. 3 frames the result in terms of the antecedent literature, offers supplementary results on the translation of the LDE result to an

---

[9] We remind the reader that Jones (1984) [30] is concerned with an economy with a finite number of consumers.

[10] It is also well known that models with a continuum of consumers and an infinite-dimensional commodity face mathematically formidable problems. These arise from the notorious failure of technical propositions such as Fatou's lemma or Lyapunov's theorem. We bypass these problems in the present paper. For the technical details, as well as their resolution, of these problems, we cite [39] and references in them.

[11] See Lemma 2.1 in [31], reproduced as Fact 5 in the Appendix below. This technique can be applied to other work; as for example, [76–78], which study economies with both indivisible and indivisible commodities in the space ℓ^{∞}, and also with and without production. This work relies on "irreducibility" assumptions on the economy and assumes that the individual endowments are in the interior of the consumption set, and we note that the techniques of this paper will also work in these settings. However, note that we are limiting ourselves to a pure exchange economy, and thereby bracketing consideration of Jones' 1984 [30] theorem on existence of equilibrium for a "private ownership economy" with a finite number of consumers and producers. The reader is also referred to [49] and [62], among others.

LIE result, and provides a discussion of its hypotheses in the light of the notions of the "overriding desirability of a divisible commodity" and of the concept of a "marginal rate of substitution"; Sects. 4 and 4.2 concern the technicalities of the proofs. This introduction already attests to a subsidiary expository motivation of the work: to provide a readable and substantively- and historically-nuanced essay on the subject rather than an epsilon extension of a known result written up in the style of a mathematical appendix.[12]

2 The Model and the Basic Existence Result

Let (K, d) be a compact metric space and $\mathcal{B}(K)$ the Borel σ-algebra on K (the smallest σ-algebra generated by open sets of K). Following [50] (1975), the economic interpretation of K is that it is a space of the commodity characteristics. A commodity bundle is defined as a (signed) measure on K, hence an element of $ca(K)$. In particular, the Dirac measure δ_t is the (one unit of) commodity bundle which contains a characteristics $t \in K$. Let $C(K)$ be the set of continuous functions on K. A net $(\boldsymbol{m}_\alpha)$ in $ca(K)$ is said to converge to a measure $\boldsymbol{m}$ in $\sigma(ca(K), C(K))$-topology or the weak* topology if and only if for every $\boldsymbol{q}(t) \in C(K)$, $\int_K \boldsymbol{q}(t) d\boldsymbol{m}_\alpha \to \int_K \boldsymbol{q}(t) d\boldsymbol{m}$.

We also includes the ℓ homogeneous consumption goods. Hence the commodity space as a whole is $\mathbb{R}^\ell \times ca(K)$. For $\boldsymbol{x} = (x^i) \in \mathbb{R}^\ell$, $\boldsymbol{x} \geq \boldsymbol{0}$ means that $x^i \geq 0$ for all $i = 1 \dots \ell$. $\boldsymbol{x} > \boldsymbol{0}$ if and only if $\boldsymbol{x} \geq \boldsymbol{0}$ and $\boldsymbol{x} \neq \boldsymbol{0}$. $\boldsymbol{x} = (x^i) \gg \boldsymbol{0}$ means that $x^i > 0$ for all $i = 1 \dots \ell$. Similarly for $\boldsymbol{m} \in ca(K)$, $\boldsymbol{m} \geq \boldsymbol{0}$ means that $\boldsymbol{m}(E) \geq 0$ for all $E \in \mathcal{B}(K)$ and $\boldsymbol{m} > \boldsymbol{0}$ if and only if $\boldsymbol{m} \geq \boldsymbol{0}$ and $\boldsymbol{m} \neq \boldsymbol{0}$. The non-negative orthant of $ca(K)$ is denoted by $\mathcal{M}(K)$, or $\mathcal{M}(K) = ca_+(K) = \{\boldsymbol{m} \in ca(K) | \ \boldsymbol{m} \geq \boldsymbol{0}\}$. $\mathcal{M}(K)$ is nothing but the set of Borel measures on K. Since K is a compact metric space, it follows from Theorems 14 and 18 of Varadarajan [80, pp. 192–194] (1965) that the space $\mathcal{M}(K)$ is a complete separable metric space in the weak* topology.

In this paper, we assume that the differentiated commodity bundles can be consumed in integer units, hence we restrict a part of the commodity space corresponding to the differentiated commodities to the set of all finite combinations of the Dirac measures with integer coefficients. Hence it is defined by the free Abelian group[13] $\mathbb{A}$ generated by the Dirac measures on K, or

$$\mathbb{A} = \{\boldsymbol{d} = \sum n_t \delta_t | \ n_t \in \mathbb{Z} \text{ and } n_t = 0 \text{ for all but finitely many } t \in K\}.$$

[12]Thus, in a mathematical appendix below, we collect and copy for the reader's convenience mathematical results available in easily-accessible but a veritable variety of sources; also see Footnotes 26 and 27 below.

[13]The free Abelian group usually appears as a basic concept for the singular homology theory in algebraic topology; see [81] for example.

The integer coefficients represent the indivisibility of the commodities. Let D be a subset of $\mathbb{A}$ which consists of finite sum of Dirac measures with non-negative coefficients.

$$D = \{\boldsymbol{d} = \sum n_t \delta_t \in \mathbb{A} |\ n_t \geq 0\},$$

hence elements of D stand for the differentiated commodities which are consumable. Then the consumption set which is assumed to be identical for each consumer is defined by

$$X = \mathbb{R}_+^\ell \times D.$$

Note that X is a complete separable metric space. From now on, we will denote $H = \mathbb{R}_+^\ell$ (homogeneous goods) and $\xi = (\boldsymbol{x}, \boldsymbol{d}) \in H \times D$ for $\xi \in X$.

We claim that D is a closed subset of $\mathcal{M}(K)$ in the weak*-topology. To see this, suppose that $\boldsymbol{d}_n \to \boldsymbol{d}$, $\boldsymbol{d}_n = \sum_{j=1}^{k_n} m_j^n \delta_{t_j} \in D$. Since $\mathcal{M}(K)$ is a metric space, we can discuss the (weak*) topology on D by the sequences rather than the nets. Since $\{\boldsymbol{d}_n\}$ is a Cauchy sequence, for $0 < \epsilon < 1$ there exists an N such that

$$\left| \int_K \mathbf{1}(t) \boldsymbol{d}_N - \int_K \mathbf{1}(t) \delta_{N+n} \boldsymbol{d}_{N+n} \right| = \left| \sum_{j=1}^{k_N} m_j^N - \sum_{j=1}^{k_{N+n}} m_j^{N+n} \right| < \epsilon$$

for all $n \geq 1$, where $\mathbf{1}(t) = 1$ for all $t \in K$. Hence $\sum_{j=1}^{k_N} m_j^N = \sum_{j=1}^{k_{N+n}} m_j^{N+n}$ for all $n \geq 1$. Therefore $\#support(\boldsymbol{d}_{N+n}) \leq \sum_{j=1}^{k_N} m_j^N < +\infty$ for all $n \geq 1$, where $\#S$ is the cardinality of a set S. Consequently we have $\#support(\boldsymbol{d}) \leq \sum_{j=1}^{k_N} m_j^N < +\infty$, or $\boldsymbol{d}$ has a finite support, $support(\boldsymbol{d}) = \{t_1 \ldots t_k\}$. We can take pairwise disjoint neighborhood U_j of t_j and write $\boldsymbol{d} = \sum_{j=1}^{k} m_j \delta_{t_j}$. Since $\boldsymbol{d}_n \to \boldsymbol{d}$, it is then easy to show that for $\boldsymbol{d}_n = \sum_{j=1}^{k_n} m_j^n \delta_{t_j^n}$, $m_j^n = m_j$, $k = k_n$ for n large enough, and $t_j^n \to t_j$, $j = 1 \ldots k$. Therefore $\boldsymbol{d} \in D$.

Set $\hat{K} = \{1 \ldots \ell\} \times K$. Then we can write $C(\hat{K}) = \{(\boldsymbol{p}, \boldsymbol{q}) |\ \boldsymbol{p} \in \mathbb{R}^\ell, \boldsymbol{q} \in C(K)\}$ and let $C_+(\hat{K}) = \{\pi = (\boldsymbol{p}, \boldsymbol{q}) \in C(\hat{K}) |\ \boldsymbol{p} \gg \mathbf{0}, \boldsymbol{q} \ggg \mathbf{0}\}$, where $\boldsymbol{q} \ggg \mathbf{0}$ means that $\boldsymbol{q}(t) \geq \epsilon$ for some $\epsilon > 0$ and for all $t \in K$. In this paper, a price vector for a commodity bundle is a nonzero element of $C(\hat{K})$. Then for $\xi = (\boldsymbol{x}, \boldsymbol{d}) \in X$ and $\pi = (\boldsymbol{p}, \boldsymbol{q}) \in C(\hat{K})$, we denote $\pi\xi = \boldsymbol{px} + \boldsymbol{qd} = \boldsymbol{px} + \int_K \boldsymbol{q}(t) d\boldsymbol{d}(t)$.

As usual, a preference relation $\succsim \subset X \times X$ is a complete, transitive and reflexive binary relation on X, and we denote $(\zeta, \xi) \notin \succsim$ by $\xi \prec \zeta$. Mas-Colell [50] (1975) made the following assumptions on the preferences and we will keep them. Let $\boldsymbol{u}_i = (0 \ldots 0, 1, 0 \ldots 0) \in \mathbb{R}^\ell$, where 1 is at the i'th place. We denote the usual topology on $\mathbb{R}^\ell$ by $\tau(\mathbb{R}^\ell)$.

Assumption (PR).

(i) $\succsim$ is closed in $X \times X$ in the (product of) $\tau(\mathbb{R}^\ell) \times \sigma(ca(K), C(K))$-topology,
(ii) for every $\xi \in X$ and every $\zeta \in X$ and , if $\xi < \zeta$, then $\xi \prec \zeta$,

(iii) for some $i = 1 \ldots \ell$ (say $i = 1$ for simplicity), and for every $\xi = (\boldsymbol{x}, \boldsymbol{d}) \in X$, there exists $\gamma > 0$ such that $\xi \prec (\gamma \boldsymbol{u}_1, \boldsymbol{0})$,

(iv) there exists $\alpha > 0$ such that for every $\xi = (\boldsymbol{x}, \boldsymbol{d}) \in X, \boldsymbol{b} \in D$ and $\rho(\boldsymbol{b}, \boldsymbol{d}) \leq 1/\alpha$, then $\xi \prec (\boldsymbol{x} + \alpha \boldsymbol{u}_i, \boldsymbol{b})$ for every $i = 1 \ldots \ell$, where ρ is a metric on X,

(v) letting $i = 1$ in *(iii)*, for every $\xi = ((x^i), \boldsymbol{d})$ and $\zeta = ((z^i), \boldsymbol{b}) \in X$, if $z^1 = 0$ and $x^1 > 0$, then $\zeta \prec \xi$.

Remark 2.1. The conditions (*i*) and (*ii*) mean the standard continuity (with respect to the weak* topology) and monotonicity assumptions, and need no further explanations. It follows from the assumption **(PR)**(*i*), nearby commodities are considered to be uniformly good substitutes. The conditions (*iii*) is the overriding desirability for the homogeneous goods. Note that the condition (*iv*) holds trivially if the set K is a finite set and/or on bounded subsets of D. Finally the condition (v) is needed even if there exist no commodity differentiation. It is related to the indivisibility of the commodities. We discuss an alternative condition in Sect. 4 which was proposed by Mas-Colell [51] (1977) to handle the indivisible commodities.

Let $\mathcal{P}$ be a subset of the set of preference relations which satisfy (*i*) through (v). Since $\mathcal{P} \subset \mathcal{F}(X \times X)$, we can endow $\mathcal{P}$ with the topology of closed convergence on $\mathcal{F}(X \times X)$, where $\mathcal{F}(X \times X)$ denotes the set of all closed subsets of a set $X \times X$; see Appendix. We assume

Assumption (PC). The set $\mathcal{P}$ is compact.

Since X is not metrizable, $\mathcal{F}(X \times X)$ is not necessarily a metric space. Jones (1983; Lemma 5) [29], however, showed that a compact subset of $\mathcal{F}(X \times X)$ is indeed metrizable. Therefore $\mathcal{P}$ is compact metric space by the assumption **(PC)**.

An initial endowment is assumed to be nonnegative vectors ω of $H \times D$, or $\omega = (\boldsymbol{e}, \boldsymbol{f}) \in H \times D$. We assume that the set Ω of all allowed endowments was a compact subset of $H \times D$ and denote it as $\Omega = \Omega_H \times \Omega_D$, $\Omega_H = [\boldsymbol{0}, \beta \boldsymbol{1}]$, where $\boldsymbol{1} = (1 \ldots 1) \in \mathbb{R}^\ell$ and $\beta > 0$ is a fixed constant, and Ω_D is an weak* compact subset of D, $\Omega_D = \{\boldsymbol{f} \in D | \ \boldsymbol{f}(K) \leq \beta\}$.

Let $i : D \to D$ be an identity map. Then $i(\boldsymbol{f}) = \boldsymbol{f}$ for every $\boldsymbol{f} \in D$. Hence we will denote $\int_{\Omega_D} i(\boldsymbol{f}) d\mu_{\Omega_D} = \int_{\Omega_D} \boldsymbol{f} d\mu_{\Omega_D}$, where the subscripts on μ denote the marginals, e.g., μ_{Ω_D} is the restriction of μ to Ω_D, and so on. The same observation applies to $\int_{\Omega_H} \boldsymbol{e} d\mu_{\Omega_H}$. We can now present further assumptions on the endowment map.

Assumption (AE). (Adequate endowments).

(i) $\int_{\Omega_H} \boldsymbol{e} d\mu_{\Omega_H} \gg \boldsymbol{0}$, and

(ii) $support\left(\int_{\Omega_D} \boldsymbol{f} d\mu_{\Omega_D}\right) = K$.

Let $(A, \mathcal{A}, \lambda)$ be a finite measure space. A map $g : A \to ca(K)$ is said to be weak*-measurable if for each $\boldsymbol{q} \in C(K)$, $\boldsymbol{q}g(a)$ is a measurable function on $(A, \mathcal{A}, \lambda)$. A weak*-measurable function $g(a)$ is said to be Gelfand integrable if there exists an element $\int_A g(a) d\lambda \in ca(K)$ such that for each $\boldsymbol{q} \in C(K)$, $\boldsymbol{q} \int_A g(a) d\lambda = \int_A \boldsymbol{q} g(a) d\lambda$ (Diestel and Uhl [16] (1977)). For every Borel set $B \in \mathcal{B}(K)$, the value of the measure $\int_A g(a)d\lambda$ at B is defined by $\int_A g(a)d\lambda(B) \equiv \int_A g(a)(B)d\lambda$. Indeed, let $\boldsymbol{q} \in C(K)$ and $\{B_i\}$ be a family of pairwise disjoint Borel sets in K. Then we have

$$\boldsymbol{q}\int_A g(a)(\cup_{i=1}^{\infty}B_i)d\lambda = \int_A \boldsymbol{q}g(a)(\cup_{i=1}^{\infty}B_i)d\lambda$$
$$= \int_A \boldsymbol{q}\sum_{i=1}^{\infty} g(a)(B_i)d\lambda = \boldsymbol{q}\sum_{i=1}^{\infty}\int_A g(a)(B_i)d\lambda,$$

hence $\int_A g(a)d\lambda(\cup_{i=1}^{\infty}B_i) = \int_A g(a)(\cup_{i=1}^{\infty}B_i)d\lambda = \sum_{i=1}^{\infty}\int_A g(a)(B_i)d\lambda = \sum_{i=1}^{\infty}\int_A g(a)d\lambda(B_i)$. The integral $\int_{\Omega_D} gd\mu_{\Omega_D}$ is considered to be a Gelfand integral of a map $i : D \to D$ over the measure space $(\Omega_D, \mathcal{B}(\Omega_D), \mu_{\Omega_D})$; see [32] for the Gelfand integral of a multifunction.

A distributionalized exchange economy is a probability measure μ on $\mathcal{M}(\mathcal{P}\times\Omega)$ and the definition of the distributional competitive equilibrium is given as follows ([22, 23] (1974, 1974)).

Definition 2.1. A pair of a price vector $\pi \in C(\hat{K})\backslash\{\mathbf{0}\}$ and a probability measure ν in $\mathcal{M}(X \times \mathcal{P} \times \Omega)$ is called a competitive equilibrium of the economy μ if the following conditions hold,

(E-1) $\nu(\{(\xi, \succsim, \omega)|\ \pi\xi = \pi\omega$ and $\xi \succsim \zeta$ whenever $\pi\zeta \leq \pi\omega\}) = 1$,
(E-2) $\int_X \xi d\nu_X \leq \int_\Omega \omega d\nu_\Omega$,
(E-3) $\nu_{\mathcal{P}\times\Omega} = \mu$, where the subscripts on ν denote the marginals (restrictions).

The main result of this paper now reads

Theorem 2.1. *There exists a competitive equilibrium* (π, ν) *with* $\pi \in C_+(\hat{K})$ *for the economy* μ *if it satisfies the assumptions* ***(PR)***, ***(PC)*** *and* ***(AE)***, *and is represented by an atomless measure space of consumers.*

We postpone the proof of this theorem to Sect. 4, and turn instead to a discussion of its substantive meaning.

3 Discussion of the Result

It is important to understand what Theorem 2.1 asserts and proves, and what it does not. It is perhaps not as widely appreciated as it ought to be that the distributionalized formulation of an exchange economy is entirely silent on the cardinality of the set of agents in the economy that is being investigated. To belabor the point, a given exchange economy represented by (say) a Dirac measure on the space of preferences and endowments may consist of a single agent, or a finite number or yet a countable or uncountable number of consumers all of the same type. The question then is how the theorem asserts the existence of a Walrasian equilibrium when the economy is interpreted so as to consist of a set of non-negligible price-takers without any convexity assumptions on their preferences. The answer of course is that the single agent in a single-agent economy, is *not* choosing his or her best bundle in the budget set but is randomizing over the attainable commodity bundles in the entire budget set. He or she is choosing lotteries! Thus Theorem 2.1, and indeed the existence results of Mas-Colell, Jones and subsequent writers working with the distributionalized formulation, connect to the type of equilibrium notions first studied by Prescott-Townsend.[14]

This section is then divided into two parts: in the first, we explore the difference between the distributionalized and individualized formulations and the extent to which Theorem 2.1 can be given an individualized representation[15]; and in the second, take the distributionalized form for granted, and examine how the theorem negotiates with the subsequent work of Mas-Colell, Jones, Yamazaki and Suzuki.

3.1 Distributionalized and Individualized Formulations

Mas-Colell (1975; Section 2.16) [50] concludes the presentation of his model and his result by a "technical comment".

> The reader familiar with the continuum of agents' models will have noted that we emphasize distributions over representations, and this in spite of the fact that for the equivalence theorem it is indispensable to consider the economy in representation form. In the present context, to deal with distributions rather than with representations is not simply a matter of taste, but it makes the existence problem a manageable one. While given an economy, we can prove the existence of an equilibrium distribution (and therefore of an equilibrium allocation), it is very unlikely that given a [particular] representation, there is an allocation which is an equilibrium with respect to [it]. Of course, the problem is not one of economic substance, it merely revolves on measurability technicalities.[16]

In the presentation of his existence theorem, Jones (1983, p. 126) [29] follows Mas-Colell's distinction between the technical and the substantive in interpreting his result.

> Notice that there is a technical problem with this interpretation. Letting $I = [0, l]$ and λ be Lebesgue measure on I, if there is a $\psi : I \to \mathcal{P} \times \Omega$ with $\mathcal{E} = \lambda \circ \psi^{-1}$, ($\psi$ will be called a representation of $\mathcal{E}$ below), it need not be true that there is an $f : I \to X$ such that $\nu = \lambda \circ (\psi, f)^{-1}$. Such an f would quite naturally be termed an assignment. However, it will be true that such an f does exist for some representation of $\mathcal{E}$. The distinction here is primarily technical. Interested readers can refer to Mas-Cole11 for further discussion.

Jones' qualification of "technical" by "primarily technical" notwithstanding, the choice over a budget set as opposed to the randomized choice over the same set, as

[14]See Precott-Townsend [66, 67]; also Forges et al. [18]. To be sure, this connection pertains only to the randomized choice of commodity bundles; these authors work with the individualized and not the distributionalized formulation of an economy.

[15]This terminology, and the question, as well as its resolutions below, is inspired by the recent results in [38] building on those in [31].

[16]We eliminate the technical symbolism from Mas-Colell's text.

well as the cardinality of the name-space of consumers, are of substantive consequence. Indeed, one may argue that this cardinality is the fundamental bifurcation of modern general equilibrium theory, distinguishing the Aumann-Vind-Brown-Robinson approach from its classical forebearers.[17] Thus from a conceptual point of view, this distributionalized approach is somewhat orthogonal to the classical theory as exposited in [2, 15, 57, 59] (1971, 1959, 1968, 2002).

The point is important enough that it is worth pursuing for the reader how this very same issue arises in non-cooperative game theory. There are two registers involved here: the first is that of finite-player, one-shot simultaneous play games, as considered in Milgrom-Weber (1985) [58]; and the other is that of large anonymous, one-shot simultaneous-play games as considered by Mas-Colell (1984a) [53].[18] Milgrom-Weber [58] consider a normal form game of incomplete information, but a distributionalized formulation is explicit in their equilibrium notion in that each player chooses, in equilibrium, a probability distribution over the product space of actions and types such that the marginal of the joint probability measure equals the given measure of types. As such, each player is choosing a behavioral strategy, one based on a randomized choice of his or her action set, rather than a pure strategy equilibrium.[19] The point is much sharper when one considers the distributionalized game in Mas-Colell [53]. Mas-Colell presents two results: the first formulates a game as a probability measure on the space of payoffs defined over a given action set, and proves the existence of a Nash equilibrium formulated as a probability distribution over the product space of actions and payoffs such that the marginal of the joint probability measure equals the given measure on payoffs, the given game, so to speak. No atomless assumptions on the given probability measure, or cardinality restrictions are required for the validity of the theorem. And so here again, we have a behavioral strategy equilibrium in that each agent, in principle is making a randomized choice over his or her action set, rather than a direct one. This makes eminent sense since there is nothing in the hypotheses of Mas-Colell's theorem that necessitates one to rule out a finite game! And it is known, at least since the time of Nash's papers that a finite game in general does not possess a pure-strategy equilibrium.

But to be sure, Mas-Colell presented [53] another result, his Theorem 2. This result required the given game to be atomless, and hence embodied an explicit restriction to a large anonymous game, as well as the action set to be finite. Under

[17]For an approach to large economies using nonstandard analysis, see Brown-Robinson (1972) [11], and the framing of the literature in Anderson (1991, 1992) [3, 4] and Khan (2008, 2012) [34, 35].

[18]The pioneering papers in these two registers are respectively Radner-Rosenthal (1982) [68] and Schmeidler (1973) [73]. The reader should also note that the original draft of [58] predates [53].

[19]For a comprehensive discussion of the technical difficulties that led Milgrom-Weber to follow the approach they did, see Khan-Rath-Sun-Yu [36]. To be sure, Milgrom-Weber also presented a pure-strategy equilibrium under the hypotheses of finite actions and conditional independence of types, a variants of results presented in Radner-Rosenthal [68]. A, by now dated, survey is available in [44].

these assumptions, Mas-Colell could show that the joint probability measure that his first theorem furnished has, as its support, the graph of a function from the space of characteristics to the space of actions. This is what Khan-Sun (1987) [41] called *symmetrization* and it has again not been as widely-understood as it ought.[20] From a substantive point of view, the assertion asserts identical players with identical payoffs take identical actions in equilibrium. And so if the game consists of a continuum of single agents each of particular type, then each agent is making a choice, not a randomized choice, in his or her action set. And so it is Theorem 2, and not Theorem 1, that furnishes a pure-strategy equilibrium.[21] But now an interesting question arises, not only in the context of the results of Milgrom-Weber [58], but more relevantly, in the context of Theorem 2.1 presented here, and the existence results of Mas-Colell and Jones presented earlier. Can the Walrasian equilibrium notions be *symmetrized* so that the joint probability distribution over preferences, endowments and commodity bundles has as its support the graph of a function from the space of preferences and endowments to the space of commodity bundles? This is to ask for an analogue of Theorem 2 of Mas-Colell [53] to the existence theorems in Mas-Colell (1975, 1977) [50, 51] and Jones (1983) [29], and of course to our Theorem 2.1 above. It is to ask whether in a Walrasian equilibrium identical types choose identical bundles, that there is no randomization over budget sets as required in the classical theory.[22] We leave this as open question and move on.

We move on to ask, in the light of recent work of [38], whether Theorem 2.1 can be parlayed into an existence theorem for the classical formulation of a large exchange economy as formulated in [5]? and whether any equilibrium distribution of a large distributionalized economy (LDE) can be represented by an equilibrium of an large individualized economy (LIE)? The answer to both questions hinges on the name-space of consumers being formalized as a *saturated* probability space. Towards explaining this notion, we recall for the reader that a finite measure space $(A, \mathcal{A}, \lambda)$ is called *essentially generated* by a family $\mathcal{G} \subset \mathcal{A}$ if the smallest σ-algebra containing $\mathcal{G}$ together with the λ-null sets is $\mathcal{A}$ itself. It is *essentially countably (uncountably) generated* if $\mathcal{G}$ is a countable (uncountable) set. For a measure space $(A, \mathcal{A}, \lambda)$, $(E, \mathcal{A}_E, \lambda_E)$ is the restriction to a subset $E \in \mathcal{A}$. We can now present:

Definition 3.1. A (finite) measure space $(A, \mathcal{A}, \lambda)$ is *saturated* (or *super-atomless*) if and only if $(E, \mathcal{A}_E, \lambda_E)$ is essentially uncountably generated for every $E \in \mathcal{A}$ with $\lambda(E) > 0$.

---

[20]See [42, 43] for this operation.

[21]See Footnote 19 for the analogy to the work in [58]. Also see [33] for a disintegration of the joint probability distribution leading, in equilibrium, to a probability measure on the space of actions corresponding to each characteristic.

[22]In the context of Milgrom-Weber [58], it is to ask whether the equilibrium distribution on the joint space of actions and types has, its support, the graph of a function from types to actions, an individualized function for each player that dictates that he or she take the same action for each of his or own revealed type, that he or she does not randomize.

Typical examples of the (homogeneous) saturated measure spaces are the atomless Loeb spaces (Loeb 1975 [48]), the product spaces of the form $[0,1]^{\kappa}$ and $\{0,1\}^{\kappa}$, where κ is an uncountable cardinal, $[0,1]$ equipped with the Lebesgue measure, and $\{0,1\}$ the "half-half" measure. The cardinalities of $[0,1]^{\kappa}$ and $\{0,1\}^{\kappa}$ are more than the continuum, however, there exists an atomless Loeb space with the continuum cardinal. Moreover, Sun-Zhang [74] and Podczeck [64] constructed a saturated measurable structure on $[0,1]$ by "enriching" the Lebesgue σ-algebra. Therefore we can assume that the measure space of consumers is saturated; for a systematic study of this assumption as applied in general equilibrium theory, see [39] and [78].

We can now define in the usual way an exchange economy $\mathcal{E}$ as a Borel measurable map $\mathcal{E} : A \to \mathcal{P} \times \Omega$, $\mathcal{E}(a) = (\succsim_a, \omega(a))$ and $\omega(a) = (\boldsymbol{e}(a), \boldsymbol{f}(a))$. A Gelfand integrable map $\xi : A \to X$ is called an allocation. Let $L^1(A, X)$ be the set of allocations. The definition of the competitive equilibrium of the individualized economy $\mathcal{E}$ is standard.

Definition 3.2. A pair of a price $\pi \in C(\hat{K})\backslash\{\mathbf{0}\}$ and an allocation $\xi \in L^1(A, X)$ is called a competitive equilibrium of the economy $\mathcal{E}$ if the following conditions hold,
(I-1) $\pi\xi(a) = \pi\omega(a)$ and if $\boldsymbol{p}\zeta \leq \boldsymbol{p}\omega(a)$, then $\xi(a) \succsim_a \zeta$ a.e.,
(I-2) $\int_A \xi(a)d\lambda = \int_A \omega(a)d\lambda$.

The assumption **(AE)** needs to be restated for the individualized economy as follows.

Assumption (AE′). (Adequate endowments).

(i) $\int_A \boldsymbol{e}(a)d\lambda \gg \mathbf{0}$, and
(ii) $support\left(\int_A \boldsymbol{f}(a)d\lambda\right) = K$.

We can now present:

Theorem 3.1. *There exists an individualized competitive equilibrium (π, ξ) with $\pi \in C_+(\hat{K})$ for the economy $\mathcal{E}$ if it is saturated and satisfies the assumptions **(PR)**, **(PC)** and **(AE′)**.*

Next, we turn to the translation of a competitive equilibrium for a distributionalized economy to one for a large individualized economy. Towards this end, let $(A, \mathcal{A}, \lambda)$ be an atomless probability measure space. For a measurable map $f : A \to \mathcal{P}\times\Omega$, recall that we are denoting the direct image measure $\lambda \circ f^{-1}$ by $f_*\lambda$.

Definition 3.3. For an economy μ, a measurable map $\mathcal{E} : A \to \mathcal{P}\times\Omega$ such that $\mu = \mathcal{E}_*\lambda$ is called a representation of μ. The representation is called saturated if the measure space $(A, \mathcal{A}, \lambda)$ is saturated.

Note that a representation is not unique even if it exists. Since $\mathcal{P} \times \Omega$ is a compact metric space, the representations of μ exists by Fact 5(i) in Appendix. Moreover, since the saturated measure spaces are atomless, saturated representations also exist. Similarly, for every allocation distribution ν, a measurable map $(\xi, \mathcal{E}) : A \to$

$X \times \mathcal{P}\times\Omega$ which satisfies $\nu = (\xi, \mathcal{E})_*\lambda$ is the representation of ν. The map $\xi : A \to X$ is nothing but an allocation. All this is simply repeating ground already traversed by Jones in the quotation above; the only new element is the availability to us of the ideas of saturation. A fundamental problem then is the following: given an equilibrium ν of an economy μ and an individualized economy $\mathcal{E}$ which represents μ, can we obtain an allocation ξ such that $(\xi, \mathcal{E})$ represents ν? The answer is generally negative if the economy is simply atomless. For the saturated economy, however, the answer is positive.

Theorem 3.2. *Let $(A, \mathcal{A}, \lambda)$ be a saturated probability space. Let distributionalized economy μ and its equilibrium ν be given. For every individualized economy $\mathcal{E} : A \to \mathcal{P}\times\Omega$ which represents μ, there exists an equilibrium allocation $\xi : A \to X$ such that $\nu = (\xi, \mathcal{E})_*\lambda$.*

3.2 Potential Alternative Hypotheses for the Result

In this subsection, we simply accept the distributionalized formulation of Theorem 2.1, and discuss the hypotheses that enable its conclusion. Put differently, we explore how in its formulation, it negotiates with the subsequent work of [29, 51, 78].

As already emphasized in the introduction, the work of Mas-Colell (1977) [51] and Jones (1983) [29] subsequent to that of Mas-Colell (1975) [50] went into two different directions, with the former eliminating differentiated commodities and the latter eliminating indivisibilities. Indivisible commodities, and more generally non-convex consumption sets as in [84], lead to the aggregate excess demand correspondence not being upper hemi-continuous; its convex-valuedness is not at issue. As such it requires assumptions additional to a multiplicity of agents. In [51], Mas-Colell proposed two conditions: the first is the assumption $\mathbf{PR}(v)$, already used above and henceforth indicated by **(ID)**, while the second is

(DD) (Dispersed divisible commodities). $\nu(a \in A|\ \boldsymbol{p}e(a) = w\}) = 0$ for every $\boldsymbol{p} = (p^i) \in \mathbb{R}^\ell_+$ with $\boldsymbol{p} \neq \mathbf{0}$ and every $w \geq 0$,[23]

In regard to **(ID)**, Mas-Colell wrote:

> The assumption ... that some amount of the divisible commodity is indispensable is very strong and shall not be postulated in our main existence theorem; it can be found in Henry; it is, however, a very convenient assumption and once the point of its dispensability is made we will use it freely.

Mas-Colell was clear as to the importance of Assumption **(DD)**.

[23]Actually, this is a generalized version of the dispersedness condition by Yamazaki (1978, 1981) [83, 84].

> [It] avoids the problem [of] consumers not switching their demands all at once (i.e., at the same price) but rather only a negligible fraction of them switching at every single price; the net effect is that mean excess demand behaves in a continuous fashion. That diversification of economic agents characteristics will induce regularity in aggregate behavior is a straightforward and old idea; it is, in fact, rather surprising that (thanks to "convexifying" results such as the Shapley-Folkman and Lyapunov theorems) equilibrium theory has managed to do without. It would appear, however, that if discrete commodities are brought into the picture, then diversification is an essential component of a reasonable theory; it is at least the most natural way to with the existence problem.[24]

Suzuki [78] relied on Assumption **(DD)** for a model in which the common consumption set is given by

$$X = \{\xi = ((x^i), (z^j)) \in \mathbb{R}^{\ell}_{+} \times \ell^{\infty} |\ 0 \leq z^j \leq M, z^j \in \mathbb{N} \text{ for all } j\},$$

where M is a positive integer. Given that **(DD)** is preferable to **(ID)** from a substantive point of view, a question arises as to why we persist in using the former as one of the hypotheses of Theorem 2.1. The reason for this is as follows. As the reader who works through the proof of Theorem 2.1 will see, the problem of the discontinuity of the aggregate excess demand potentially arises on the set of consumers $\{a \in A|\ \pi(\mathbf{0}, z) = \pi\omega(a)\}$, where $z = (z^j)$ and $z^j (\leq M) \in \mathbb{N}$. The condition **(ID)** was used to avoid the case $\zeta = (\mathbf{0}, z)$ directly. The condition **(DD)** was applicable for the space ℓ^{∞} since in this case we have

$$\begin{aligned} \nu(\{a \in A|\ \pi(\mathbf{0}, z) = \pi\omega(a)\}) &= \nu\left(\left\{a \in A \,\middle|\, \sum_{i=1}^{\ell} p^i e^i(a) = \sum_{j=1}^{\infty} q^j (z^j - f^j(a))\right\}\right) \\ &= 0, \end{aligned}$$

where $\pi = ((p^i), (q^j)) \in \mathbb{R}^{\ell}_{+} \times \ell^1_{+}$, $\omega(a) = ((e^i(a)), (f^j(a))) \in \mathbb{R}^{\ell}_{+} \times \ell^{\infty}_{+}$. The last equality follows since the set $\{\sum_{j=1}^{\infty} q^j (z^j - f^j(a))\}$ is a countable set under the assumption that $\{(f^j(a))|\ a \in A\}$ is also a countable set.

For the space $ca(K)$, the situation is entirely different. Write $\boldsymbol{d} = \sum_{j=1}^{m} d_j \delta_{t_j}$, $\boldsymbol{e}(a) = (e^i(a))$ and $\boldsymbol{f}(a) = \sum_{k=1}^{n} f_k(a)\delta_{t_k}$. What we have to show is

$$\nu\left(\left\{a \in A \,\middle|\, \sum_{i=1}^{\ell} p^i e^i(a) = \sum_{j=1}^{m} \boldsymbol{q}(t_j) d_j - \sum_{k=1}^{n} \boldsymbol{q}(t_k) f_k(a)\right\}\right) = 0,$$

but since t_j and t_k take their values in an uncountable set K, the set $\{\sum_{j=1}^{m} \boldsymbol{q}(t_j) d_j - \sum_{k=1}^{n} \boldsymbol{q}(t_k) f_k(a)\}$ is obviously an uncountable set even if the range $\{\boldsymbol{f}(a)|\ a \in A\}$ is assumed to be countable.

---

[24] We have routinely modified Mas-Colell's prose to fit the continuity of our text. It is also worth mentioning that Mas-Colell credits Walras for this idea.

Next, we turn to Jones [29] who works with the consumption set $X = \mathbb{R}^{\ell}_{+} \times \mathcal{M}(K)$, we denote $(\boldsymbol{x}, \boldsymbol{m}) \in X, \boldsymbol{x} \in \mathbb{R}^{\ell}_{+}, \boldsymbol{m} \in \mathcal{M}(K)$.[25] Note that a measure $\boldsymbol{m}$ can take any non-negative (real) values. Instead of indivisibility, Jones' condition, in the revised version due to [60] is given as:

(UHS) For all $\gamma > 1$, there is a $\eta > 0$ such that for all $\alpha > 0$ and for all $\boldsymbol{m} \in \mathcal{M}(K)$ and $\succsim \in \mathcal{P}$, if $d(s,t) < \eta$, then $\boldsymbol{m} + \alpha\delta_s \prec \boldsymbol{m} + \alpha\gamma\delta_t$.

Intuitively, the condition **(UHS)** says that the marginal rates of substitutions between nearby commodities are uniformly (with respect to ξ and $\succsim$) close to one. The crucial proposition for the proof of Theorem 2.1 is Lemma 4.3 in which we show the equilibrium prices $\boldsymbol{q}_n$ for an approximating economy with a finite set of characteristics K_n, are equi-continuous. The point is that we can prove this lemma by relying on **(UHS)** instead of the indivisibility of the commodities.

We briefly indicate the argument. Suppose that $(K_n, \boldsymbol{q}_n)$ are not equi-continuous. Then we can assume that there exist sequences (t_n), (s_n) such that

$$d(t_n, s_n) \to 0 \text{ and } \lim_{n\to\infty} \frac{\boldsymbol{q}_n(t_n)}{\boldsymbol{q}_n(s_n)} > 1.$$

As in Lemma 4.3, for any $t, s \in K$ and $\boldsymbol{m} \in \mathcal{M}(K)$ with $\boldsymbol{m}(\{t\}) > 0$, we define $\boldsymbol{m}^t_s = \boldsymbol{m} - \boldsymbol{m}(\{t\})\delta_t + (\boldsymbol{q}_n(t)/\boldsymbol{q}_n(s))\boldsymbol{m}(\{t\})\delta_s$. Then as in the proof of Lemma 4.3, we can show that for each n,

$$\lambda\left(\{a \in A \mid \boldsymbol{m}_n(a)(\{t_n\}) > 0\}\right) > 0,$$

hence it follows from **(UHS)** that for n sufficiently large

$$(\boldsymbol{x}_n(a), \boldsymbol{m}_n(a)) \prec_a (\boldsymbol{x}_n(a), \boldsymbol{m}_{n\,s_n}^{\,t_n}(a)).$$

This is a contradiction, since $\boldsymbol{q}_n \boldsymbol{m}_{n\,s_n}^{\,t_n}(a) = \boldsymbol{q}_n z_n(a)$.

We leave it to the reader to check that the validity of the conclusions of Lemmata 4.3 and 4.6 below do not require **(PR)**(v) when there are no indivisible commodities.

4 Proofs of Theorems

In keeping with the subsidiary expository motivation of the essay, rather than sketches, we present the proofs in detail. Indeed, Lemmata 4.2 to 4.6 constitute the technical heart of the paper, one encapsulating almost the entire body of the proof.

[25] In fact, Jones did not assume that the commodity space was the product of $\mathbb{R}^{\ell}$ and $\mathcal{M}(K)$. But for notational consistency between the text, we keep this setting.

Their clear presentation is also necessary from a substantive point of view as they put into juxtaposition the twin conditions for handling the indivisible commodities (the former part of Sect. 3.2), and the twin assumptions of indivisible commodities and the bounded marginal rates of substitution[26] (the latter part of Sect. 3.2). For this purpose, Lemmata 4.2 and 4.6 are needed to be presented in full, and consequently yield our final outcome.

4.1 Proof of Theorem 2.1

The idea of proof of Theorem 2.1 follows that of Mas-Colell [50] or Jones [29]. Let $(A, \mathcal{A}, \lambda)$ be a complete and atomless probability space. Then by Fact 5(i) in Appendix, there exists a Borel measurable map $\mathcal{E} : A \to \mathcal{P} \times \Omega$ such that $\mu = \mathcal{E}_{*}\lambda$. The (mapping) economy $\mathcal{E}$ is nothing but a representation of the economy μ.

Take a sequence ϵ_n of positive numbers decreasing to zero. As in [29, 50], we can construct a sequence of finite subsets $K_n = \{t_n^1 \dots t_n^{m_n}\}$ of K and a sequence of pairwise disjoint open sets B_n^i with $t_n^i \in B_n^i$ for $i = 1 \dots m_n$ such that denoting $B_n = \cup_{i=1}^{m_n} B_n^i$, $d(t^i, t) \leq \epsilon_n$ for every $t \in B_n^i$ and for all n, $i = 1 \dots m_n$,

$$\int_A f(a)d\lambda(B_n) = \int_A f(a)d\lambda(K) \text{ and } K_n \subset K_{n+1} \text{ for all } n,$$

and $K_n \to K$ in the topology of closed convergence.

Let $L_n = LS(t_n^1 \dots t_n^{m_n}) \subset ca(K)$ be the linear space spanned by $\{\delta_{t_n^1} \dots \delta_{t_n^{m_n}}\}$ and set $D_n = D \cap L_n$, $n = 1, 2 \dots$. We then define

$$X_n = H \times D_n, \quad \Omega_n = \Omega \cap (H \times D_n), \quad \mathcal{P}_n = \mathcal{P} \cap \mathcal{F}(X_n \times X_n).$$

Let r_n be a map from $\mathcal{P}$ to $\mathcal{P}_n$ defined by $r_n(\succsim) = \succsim^n = \succsim \cap (X_n \times X_n)$, $n = 1, 2 \dots$. Jones [29, Lemma 6] proved that r_n is continuous, and it is obvious that $r_n(\succsim) \to \succsim$ as $n \to +\infty$. For each n, let $\psi_n : \mathcal{P} \times \Omega \to \mathcal{P}_n \times \Omega_n$ be a map defined by

$$\psi_n(\succsim, \omega) = (\succsim^n, \omega_n), \quad \succsim^n = r_n(\succsim), \quad \omega_n = (\boldsymbol{e}_n, \boldsymbol{f}_n),$$

$$\boldsymbol{e}_n = \boldsymbol{e}, \ \boldsymbol{f}_n = \sum_{j=1}^{m_n} \boldsymbol{f}(B_n^j)\delta_{t_n^j}.$$

Since r_n is continuous and B_n^i are open, the map ψ_n is measurable. Set $\mathcal{E}_n = \psi_n \circ \mathcal{E} : A \to \mathcal{P}_n \times \Omega_n$, $\mathcal{E}_n(a) = (\succsim_a^n, \omega_n(a))$. Then $\mathcal{E}_n$ is a mapping economy with finite number of commodities under the identification X_n with $\mathbb{R}_+^{\ell+m_n}$. We have by construction $\int_A \boldsymbol{e}_n(a)d\lambda \gg \mathbf{0}$, and $\boldsymbol{f}_n(a)(K) = \boldsymbol{f}(a)(K)$, $n = 1, 2 \dots$.

[26] Lemma 4.2 is especially crucial in this regard. For this emphasis on exposition, also see Footnotes 12 and 27, and the text they footnote, above.

Lemma 4.1. *The (mapping) economy $\mathcal{E}_n$ has a quasi-equilibrium, or there exist a price vector $\pi_n \in \mathbb{R}^{\ell}_{+} \times \mathbb{R}^{m_n}_{+}$ with $\pi \neq \mathbf{0}$ and an allocation $\xi_n : A \to X_n$, $\xi_n(a) = (\boldsymbol{x}_n(a), \boldsymbol{d}_n(a))$ which satisfy*
$\pi_n \xi_n(a) = \pi_n \omega_n(a)$ and $\xi_n(a) \succsim^n_a \zeta$ whenever $\pi_n \zeta < \pi_n \omega_n(a)$ a.e. in A,
$\int_A \xi_n(a) d\lambda \leq \int_A \omega_n(a) d\lambda$.

Proof. See Khan-Yamazaki [45, Proposition 2].

Without loss of generality, we can assume $\|\pi_n\| = max\{p^i_n, \boldsymbol{q}_n(t) | \ i = 1 \ldots \ell, t \in K_n\} = 1$ for all n, and we will identify the vector $\boldsymbol{q}_n \in \mathbb{R}^{m_n}_{+}$ with a function on K_n which is still denoted as $\boldsymbol{q}_n$. Let $(K_n, \boldsymbol{q}_n)$ be a sequence as above. We will say that it is *equi-continuous* if for all $\epsilon > 0$ there is a $\delta > 0$ such that for all $t, s \in K_n$ with $d(t, s) \leq \delta$, $|\boldsymbol{q}_n(t) - \boldsymbol{q}_n(s)| \leq \epsilon$. In the next lemma, the assumption of indivisible commodities plays an essential role.

Lemma 4.2. *$(K_n, \boldsymbol{q}_n)$ are equi-continuous.*

Proof. For any $t, s \in K$ and $\boldsymbol{d} \in D$ with $\boldsymbol{d}(t) > 0$, we define $\boldsymbol{d}^t_s = \boldsymbol{d} - \delta_t + \delta_s$. That is, $\boldsymbol{d}^t_s$ is the bundle which is obtained from $\boldsymbol{d}$ by replacing one unit of t by one unit of s. Since K is compact, it follows from the assumption **(PR)**(iv) that for every $t, s \in K$, $\boldsymbol{x} \in \mathbb{R}^{\ell}_{+}$ and $\succsim \in \mathcal{P}$,

$$d(t, s) < 1/\alpha \ \text{ implies } \ (\boldsymbol{x}, \boldsymbol{d}) \prec (\boldsymbol{x} + \alpha \boldsymbol{u}_i, \boldsymbol{d}^t_s), \ \ i = 1 \ldots \ell.$$

Suppose that $(K_n, \boldsymbol{q}_n)$ are not equi-continuous. Then we can assume that there exist sequences (t_n), (s_n) such that

$$d(t_n, s_n) \to 0 \ \text{ and } \ \lim_{n \to \infty} \boldsymbol{q}_n(t_n) > \lim_{n \to \infty} \boldsymbol{q}_n(s_n) + \theta \ \text{ for some } \ \theta > 0.$$

Since $\int_A \boldsymbol{f}(a) d\lambda \gg \mathbf{0}$, one obtains that for each n,

$$\lambda \left(\{a \in A | \ \boldsymbol{d}_n(a)(t_n) > 0\}\right) > 0.$$

Then it follows for n sufficiently large that

$$(\boldsymbol{x}_n(a), \boldsymbol{d}_n(a)) \prec_a (\boldsymbol{x}_n(a) + (\theta/2)\boldsymbol{u}_i, (\boldsymbol{d}_n)^{t_s}_{s_n}(a)).$$

This is a contradiction, since

$$\begin{aligned}
(\boldsymbol{p}_n, \boldsymbol{q}_n)&(\boldsymbol{x}_n(a) + (\theta/2)\boldsymbol{u}_i, (\boldsymbol{d}_n)^{t_n}_{s_n}(a)) \\
&= \boldsymbol{p}_n \boldsymbol{x}_n(a) + (\theta/2)p^i_n + \boldsymbol{q}_n \boldsymbol{d}_n(a) - \boldsymbol{q}_n(t_n) + \boldsymbol{q}_n(s_n) \\
&\leq \boldsymbol{p}_n \boldsymbol{x}_n(a) + \theta/2 + \boldsymbol{q}_n \boldsymbol{d}_n(a) - \theta \\
&= (\boldsymbol{p}_n, \boldsymbol{q}_n)(\boldsymbol{x}_n(a), \boldsymbol{d}_n(a)) - \theta/2 \leq (\boldsymbol{p}_n, \boldsymbol{q}_n)(\boldsymbol{e}_n(a), \boldsymbol{f}_n(a)).
\end{aligned}$$

□

Since the sequence $\{\boldsymbol{q}_n\}$ is uniformly bounded, it follows from Fact 9 in the Appendix that we can assume that $(K_n, \boldsymbol{q}_n) \to (K, \boldsymbol{q})$ for some $\boldsymbol{q} \in C(K)$. We can now assume that $\boldsymbol{p}_n \to \boldsymbol{p}$ and $\boldsymbol{x}_n(a) \to \boldsymbol{x}(a)$ by Fatou's Lemma (Fact 4 in Appendix). The next lemma states that the equilibrium prices are uniformly bounded away from 0.

Lemma 4.3. *If $\pi\omega(a) > 0$ and $\boldsymbol{q}(t^*) = 0$ for some $t^* \in K$, then $\boldsymbol{d}_n(a)(K) \to +\infty$.*

Proof. Suppose not. Then by taking subsequence if necessary, we can assume that $\boldsymbol{d}_n(a) \to \boldsymbol{d}^*(a)$. First we shall show that $p^1 > 0$. Suppose not. Then by the assumption **(PR)**(iii), there exists a $\gamma > 0$ satisfying that $(\boldsymbol{x}(a), \boldsymbol{d}^*(a)) \prec_a (\gamma\boldsymbol{u}_1, \boldsymbol{0})$. Since $\pi\omega(a) > 0$, certainly we have $\boldsymbol{p}(\gamma\boldsymbol{u}_1) < \pi\omega(a)$, hence $(\boldsymbol{x}_n(a), \boldsymbol{d}_n(a)) \prec_a (\gamma\boldsymbol{u}_1, \boldsymbol{0})$ and $\boldsymbol{p}_n(\gamma\boldsymbol{u}_1) < \pi_n\omega_n(a)$ for n sufficiently large, since $\pi_n\omega_n(a) \to \pi\omega(a)$ by Fact 8 in Appendix. A contradiction.

We claim that $(\boldsymbol{x}(a), \boldsymbol{d}^*(a)) \precsim_a (\boldsymbol{x}, \boldsymbol{d}) = ((x^i), \boldsymbol{d})$ whenever $\boldsymbol{px} + \boldsymbol{qd} \leq \pi\omega(a)$. To see this, suppose that $(\boldsymbol{x}(a), \boldsymbol{d}^*(a)) \prec_a (\boldsymbol{x}, \boldsymbol{d})$ and $\boldsymbol{px} + \boldsymbol{qd} \leq \pi\omega(a)$. We can assume that $x^1 > 0$, since otherwise $(\boldsymbol{x}(a), \boldsymbol{d}^*(a)) \precsim_a (\boldsymbol{x}, \boldsymbol{d})$ by the assumption **(PR)**(v). Hence we can assume without loss of generality that $\boldsymbol{px} + \boldsymbol{qd} < \pi\omega(a)$, since $\pi\omega(a) > 0$ and $p^1 > 0$. Then we have a sequence $(\boldsymbol{x}_n, \boldsymbol{d}_n)$ with $(\boldsymbol{x}_n, \boldsymbol{d}_n) \to (\boldsymbol{x}, \boldsymbol{d})$ and $\boldsymbol{p}_n\boldsymbol{x}_n + \boldsymbol{q}_n\boldsymbol{d}_n < \pi_n\omega_n(a)$, since $\pi_n\omega_n(a) \to \pi\omega(a)$ by Fact 8 in Appendix. Hence $(\boldsymbol{x}_n(a), \boldsymbol{d}_n(a)) \prec_a (\boldsymbol{x}_n, \boldsymbol{d}_n)$ and $\boldsymbol{p}_n\boldsymbol{x}_n + \boldsymbol{q}_n\boldsymbol{d}_n < \pi_n\omega_n(a)$ for n large enough, a contradiction.

But this can not be the case, since $\boldsymbol{px}(a) + \boldsymbol{q}(\boldsymbol{d}^*(a) + \delta_{t*}) \leq \pi\omega(a)$ and $(\boldsymbol{x}(a), \boldsymbol{d}^*(a)) \prec_a (\boldsymbol{x}(a), \boldsymbol{d}^*(a) + \delta_{t*})$. Therefore $\boldsymbol{d}_n(a)(K) \to +\infty$.

Lemma 4.4. *There exists an $\hat{\delta} > 0$ with $\boldsymbol{q}_n(t) \geq \hat{\delta}$ for all n and for all $t \in K$.*

Proof. If the lemma was false, there exists a sequence $t_n \in K_n$ with $t_n \to t^* \in K$ and $\boldsymbol{q}_n(t_n) \to 0$. Then we have $\boldsymbol{q}(t^*) = 0$. By the assumption **(AE)** there exists a measurable subset U of A with positive measure such that $\pi\omega(a) > 0$ on U. Then we have $\int_A \boldsymbol{f}_n(a)(K)d\lambda = \int_A \boldsymbol{d}_n(a)(K)d\lambda \geq \int_U \boldsymbol{d}_n(a)(K)d\lambda \to +\infty$ from Lemma 4.3, and this contradicts that
$\int_A \boldsymbol{f}_n(a)(K)d\lambda \to \int_A \boldsymbol{f}(a)(K)d\lambda < +\infty$.

Take an $\epsilon > 0$. Then it follows from Lemma 4.4 that

$$\hat{\delta}\boldsymbol{d}_n(a)(K) \leq \boldsymbol{q}_n\boldsymbol{d}_n(a) \leq \pi_n\omega_n(a) < \|\omega(a)\| + \epsilon,$$

therefore $0 \leq \|\boldsymbol{d}_n(a)\| < (\|\omega(a)\| + \epsilon)/\hat{\delta} \leq (\beta + \epsilon)/\hat{\delta}$ for all n large enough a.e. Let $\hat{D} = \{\boldsymbol{d} \in D \mid \boldsymbol{d} \leq (\beta+\epsilon)/\hat{\delta}\}$ and $\hat{X} = H \times \hat{D}$. $\hat{D}_n$ and $\hat{X}_n$ are defined in a similar way. Note that $\hat{D}_n$ and $\hat{D}$ are compact metric spaces, and $\hat{X}$ and $\hat{X}_n$ are locally compact metric spaces. Then we can define the distributional equilibrium (π_n, ν_n) where ν_n is a probability measure on $\hat{X}_n \times \mathcal{P}_n \times \Omega_n$ defined by $\nu_n = \lambda_*(\boldsymbol{x}_n, \boldsymbol{d}_n, \precsim^n, \omega_n)$. The measure ν_n can be naturally extended to a measure on $X \times \mathcal{P} \times \Omega$ defined by $\nu_n(B) = \nu_n(B \cap (\hat{X}_n \times \mathcal{P}_n \times \Omega_n))$ for every $B \in \mathcal{B}(X \times \mathcal{P} \times \Omega)$. Let $\eta_1^n = \lambda_*\boldsymbol{x}_n$ and $\eta_2^n = \lambda_*(\boldsymbol{d}_n, \precsim^n, \omega_n)$. By Fact 4 in Appendix, we have $\boldsymbol{x}_n(a) \to \boldsymbol{x}(a)$ a.e., hence it follows from Fact 5(i) in Appendix that $\eta_1^n \to \eta_1$. By Fact 3 in Appendix, we have $\eta_2^n \to \eta_2$. Therefore it follows from Fact 5(*ii*) in Appendix that $\nu_n \to \nu$ with $\nu_H = \eta_1$ and $\nu_{\hat{D}\times\mathcal{P}\times\Omega} = \eta_2$. Obviously $\nu_{\mathcal{P}\times\Omega} = \mu$, hence the condition (E-3) of Definition 2.1 is met.

We will show that (π, ν) is an equilibrium for $\mathcal{E}$. We have $\int_X \xi d\nu_n - \int_\Omega \omega d\nu_n = \int_A \xi_n(a)d\lambda - \int_A \omega_n(a)d\lambda \leq \mathbf{0}$ for all n, hence $\int_X \xi d\nu - \int_\Omega \omega d\nu \leq \mathbf{0}$ by the change of variable formula. This proves the condition (E-2) of Definition 2.1.

Lemma 4.5. $\nu_{X\times\Omega}(\{(\xi,\omega)|\ \pi\xi = \pi\omega\}) = 1.$

Proof. Define $E = \{(\xi,\omega) \in \hat{X} \times \Omega|\ \pi\xi = \pi\omega\}$ and $E_n = \{(\xi_n,\omega_n) \in \hat{X} \times \Omega|\ \pi_n\xi_n = \pi_n\omega_n\}$. We shall show that $\nu_{\hat{X}\times\Omega}(E) = 1$, hence $\nu_{X\times\Omega}(E) = 1$. To this end it suffices to show that $Ls(E_n) \subset E$, where Ls denotes the topological limes superior [26, p. 15] (1974). Indeed, since $\mathcal{F}(\hat{X} \times \Omega)$ is a compact metric space, there exists a converging subsequence E_{n_k} of E_n with $E_{n_k} \to \hat{E}$ in the topology of closed convergence. Since $\hat{X} \times \Omega$ is a locally compact and σ-compact metric space, it follows from Fact 7 in Appendix that $1 = \nu_{X\times\Omega}(\hat{E}) = \nu_{X\times\Omega}(Ls(E_{n_k})) \leq \nu_{X\times\Omega}(Ls(E_n)) \leq 1$. Let $(\xi,\omega) \in Ls(E_n)$ and take a sequence $(\xi_n,\omega_n) \in Ls(E_n)$ with $(\xi_n,\omega_n) \to (\xi,\omega)$. $(\xi,\omega) \in E$ immediately follows from Fact 8 in Appendix. □

The proof is completed by the next lemma.

Lemma 4.6. $\nu(\{(\xi,\precsim,\omega)|\xi \succsim \zeta \textit{ whenever } \pi\zeta \leq \pi\omega\}) = 1.$

Proof. The proof proceeds in the same way as that of Lemma 4.5. Define

$$F_n = \{(\xi_n,\precsim^n,\omega_n) \in \hat{X} \times \mathcal{P} \times \Omega|\ \xi_n \succsim^n \zeta_n \text{ whenever } \pi_n\zeta_n \leq \pi_n\omega_n\},$$

$$F = \{(\xi,\precsim,\omega) \in \hat{X} \times \mathcal{P} \times \Omega|\ \xi \succsim \zeta \text{ whenever } \pi\zeta \leq \pi\omega\}.$$

We shall show that $\nu_{\hat{X}\times\mathcal{P}\times\Omega}(F) = 1$, hence $\nu_{X\times\mathcal{P}\times\Omega}(F) = 1$. It suffices to show that $Ls(F_n) \subset F$. Let $(\xi,\precsim,\omega) \in Ls(F_n)$ and take a sequence $(\xi_n,\precsim^n,\omega_n) \in Ls(F_n)$ such that $(\xi_n,\precsim^n,\omega_n) \to (\xi,\precsim,\omega)$. We want to show $(\xi,\precsim,\omega) \in F$.

Suppose that there exists $\zeta \in X$ such that $\pi\zeta \leq \pi\omega$ and $\xi \prec \zeta$. We can assume without loss of generality that $\zeta = (\boldsymbol{z},\boldsymbol{b}) = ((z^i),\boldsymbol{b}) \in H\times D, z^1 > 0$. Indeed, if $z^1 = 0$, then $\xi \succsim \zeta$ by the assumption **(PR)**(v). If $\pi\omega > 0$, then since the preferences are continuous and $p^1 > 0$ (see Lemma 4.3), we can assume that $\pi\zeta < \pi\omega$ and $\xi \prec \zeta$. This contradicts $(\xi_n,\precsim^n,\omega_n) \in Ls(f_n)$, since $\pi_n\zeta_n < \pi_n\omega_n$ and $\xi_n \prec \zeta_n$ for n large enough, where ζ_n is defined in the similar way as ξ_n. Therefore we have shown that $\xi \succsim \zeta$ whenever $\pi\zeta \leq \pi\omega$ and $\pi\omega > 0$. Then by the assumptions **(AE)** and **(PR)**(ii), we have $\boldsymbol{p} \gg \mathbf{0}$.

For ω with $\pi\omega = 0$, then since $\boldsymbol{p} \gg \mathbf{0}$ and $\boldsymbol{q}(t) \geq \hat{\delta}$ for all $t \in K$, we have $\{\xi \in X|\ \pi\xi \leq \pi\omega = 0\} = \{\mathbf{0}\}$, hence $\xi = \mathbf{0}$ is trivially a maximal element in the budget set, hence $(\xi,\precsim,\omega) \in F$. □

4.2 Proofs of Theorems 3.1 and 3.2

First we need to show

Lemma 4.7. *Let $(A, \mathcal{A}, \lambda)$ be an atomless measure space, $\mathcal{E} : A \to \mathcal{P} \times \Omega$ be a representation of μ and $\xi : A \to X$ a measurable mapping. Define $\nu = (\xi, \mathcal{E})_* \lambda$. Then ξ is an equilibrium allocation of $\mathcal{E}$ if and only if ν is an equilibrium distribution of μ.*

Proof. Suppose that ξ is an equilibrium allocation of $\mathcal{E}$. Then there exists a price vector $\pi(\neq \mathbf{0}) \in C_+(\hat{K})$ with $\lambda(G) = 1$, and $\int_A \xi(a)d\lambda = \int_A \omega(a)d\lambda$, where $G = \{a \in A | \ \pi\xi(a) = \pi\omega(a)$ and $\xi(a) \succsim_a \zeta$ whenever $\pi\zeta \leq \pi\omega(a)\}$. Let $H = \{(\xi, \succsim, \omega) \in X \times \mathcal{P} \times \Omega | \ \pi\xi = \pi\omega$ and $\xi \succsim \zeta$ whenever $\pi\zeta \leq \pi\omega\}$. Then $(\xi, \mathcal{E})(G) = H$, hence $\nu(H) = (\xi, \mathcal{E})_*\lambda(H) = \lambda(G) = 1$, which proves the condition (E-1). Since $\xi_*\lambda = \nu_X$ and $\omega_*\lambda = \nu_\Omega = \mu_\Omega$, the change of variable formula gives $\int_X \xi d\nu_X = \int_\Omega \omega d\mu_\Omega$. Hence the condition (E-2) is verified. Finally, the condition (E-3) follows from $\nu_{\mathcal{P}\times\Omega} = \mathcal{E}_*\lambda = \mu$. The converse is also proved in a similar way. □

Let $\mathcal{E}$ be a saturated economy which satisfies all assumptions of Theorem 3.1 and set $\mu = \mathcal{E}_*\lambda$. Then the distributionalized economy μ satisfies all of the assumption of Theorem 2.1, hence there exists an equilibrium (π, ν). Since $(A, \mathcal{A}, \lambda)$ is saturated, there exists a measurable map $\xi : A \to X$ such that $(\xi, \mathcal{E})_*\lambda = \nu$ by Fact 6 in Appendix. Then by Lemma 4.7, (π, ξ) is an equilibrium of $\mathcal{E}$. This proves Theorem 3.1. Finally Theorem 3.2 follows immediately from Fact 6 in Appendix and Lemma 4.7. □

Appendix

In this section, we collect some mathematical results for the readers convenience.[27] The following is easy to see from [16, pp. 53–54].

Fact 1. *If $g : A \to ca(K)$ is weak* measurable and $\mathbf{q}g(a)$ is integrable function for all $\mathbf{q} \in C(K)$, then g is Gelfand integrable.*

It is easy to prove

Fact 2. *Let $\{g_n\}$ be a sequence of Gelfand integrable functions from A to $ca(K)$ which converges a.e. to g in the weak* topology. Then it follows that $\int_A g_n(a)d\lambda \to \int_A g(a)d\lambda$ in the weak* topology.*

The next fact is well known; see for example [71, Chapter 14].

[27] This is in keeping with the expository intent of the essay; see Footnotes 12 and 26 and the text they footnote.

Fact 3. *Let K be a compact metric space. The weak* topology of $ca(K)$ is separable and norm bounded subsets of $ca(K)$ are compact and metrizable, in particular, the set of probability measures on K is compact with respect to the weak* topology.*

The next fact is also well known as Fatou's lemma in ℓ dimensions; see [26, p. 69].

Fact 4. *Let $(g_n)_{n \in \mathbb{N}}$ be a sequence of integrable functions of a measure space $(A, \mathcal{A}, \lambda)$ to $\mathbb{R}^\ell_+$. Suppose that $\lim_n \int_A g_n(a)d\lambda$ exists. Then there exists an integrable function $g : (A, \mathcal{A}, \lambda) \to \mathbb{R}^\ell_+$ such that*

(*a*) $g(a) \in Ls(g_n(a))$ *a.e. in* A
(*b*) $\int_A g(a)d\lambda \leq \lim_{n\to\infty} \int_A g_n(a)d\lambda$.

Let (Y, d) be a complete separable metric space (Polish space), $(A, \mathcal{A}, \lambda)$ a complete and atomless probability space and g a Borel measurable map from $(A, \mathcal{A}, \lambda)$ to $ca(Y)$. The direct image measure $\lambda \circ g^{-1}$ is denoted by $g_*\lambda$. The operator $_*\lambda$ is a map from the set of all measurable maps of A to Y which is denoted by $L^0(A, Y)$ to $\mathcal{M}(Y)$ where $L^0(A, Y)$ is the set of all (Borel) measurable functions from A to Y with the topology of convergence in measure. Recall that if a sequence of measurable functions converges a.e., it converges in measure. We then have from [31, Lemma 2.1]

Fact 5.

(*i*) *The map $_*\lambda : L^0(A, Y) \to \mathcal{M}(Y)$ is surjective and continuous.*
(*ii*) *Let $(g_n)_{n\in\mathbb{N}}$ be a sequences in $L^0(A, Y)$ and $(h_n)_{n\in\mathbb{N}}$ a sequences in $L^0(A, Z)$, where Y and Z are Polish spaces. Suppose that $\lambda_* g_n \to \eta_1 \in \mathcal{M}(Y)$ and $\lambda_* h_n \to \eta_2 \in \mathcal{M}(Z)$. Then some sub-sequence of $\lambda_*(g_n, h_n)$ converges (weakly) to $\theta \in \mathcal{M}(Y \times Z)$ such that $\theta_Y = \eta_1$ and $\theta_Z = \eta_2$, respectively.*

The next fact reveals a striking property of the saturated space identified in [31, Definition 2.2].

Fact 6. *Let Z and Y are complete separable metric spaces. Then a finite measure space $(A, \mathcal{A}, \lambda)$ is saturated if and only if for every measure $\nu \in \mathcal{M}(Z \times Y)$ and measurable function $\mathcal{E} : A \to Y$ with $\mathcal{E}_*\lambda = \nu_Y$, there exists a measurable function $\xi : A \to Z$ which satisfies $(\xi, \mathcal{E})_*\lambda = \nu$.*

Let Z be a Hausdorff topological space. We denote the set of all closed subsets of a set Z by $\mathcal{F}(Z)$. The topology τ_c on $\mathcal{F}(Z)$ of closed convergence is a topology which is generated by the base

$$[C; G_1 \dots G_n] = \{f \in \mathcal{F}(Z) |\ F \cap C = \emptyset, F \cap G_i \neq \emptyset, i = 1 \dots n\}$$

as C ranges over the compact subsets of Z and G_i are arbitrarily finitely many open subsets of Z. It is well known that if Z is locally compact, the space $\mathcal{F}(Z)$ is a compact metric space; see for example [26, pp. 15–19]. Recall that a set $K \subset Z$ is called σ-compact if it is a countable union of compact subsets of Z.

Fact 7. *If F_n is a sequence of closed subsets of a locally compact and σ-compact metric space K such that $F_n \to F$ in the topology of closed convergence and μ_n is a sequence of probability measures on K such that $\mu_n(f_n) = 1$ for all n and $\mu_n \to \mu$, then $\mu(f) = 1$.*

Proof. The proof is similar to that available in [76, Fact 5]. □

Let $\{K_n\}$ be an increasing sequence of closed subsets of a compact metric space K converging to K in the topology of closed convergence. If $\boldsymbol{q}_n : K_n \to \mathbb{R}$ is continuous, we will write $(K_n, \boldsymbol{q}_n) \to (K, \boldsymbol{q})$ if $\boldsymbol{q} \in C(K)$ and for every subsequence n_k and $t_{n_k} \in K_{n_k}$ with $t_{n_k} \to t$, $\boldsymbol{q}(t_{n_k}) \to \boldsymbol{q}(t)$. We have from [50] (1975)

Fact 8. *Let $\{\boldsymbol{m}_n\}$ be a bounded sequence in $ca(K)$ with $support(\boldsymbol{m}_n) \subset K_n$, and $(K_n, \boldsymbol{q}_n) \to (K, \boldsymbol{q})$. Then $\boldsymbol{q}_n \boldsymbol{m}_n \to \boldsymbol{q}\boldsymbol{m}$.*

Let $(K_n, \boldsymbol{q}_n)$ be a sequence as above. We will say that it is equi-continuous if for all $\epsilon > 0$ there is a $\delta > 0$ such that for all $t, s \in K_n$ with $d(t, s) \leq \delta$, $|\boldsymbol{q}_n(t) - \boldsymbol{q}_n(s)| \leq \epsilon$. The following is available in [50].

Fact 9. *Let $\{K_n\}$ be a sequence of closed sets of a compact metric space K with $K_n \subset K_{n+1} \subset \cdots \to K$ in the topology of closed convergence and $\{\boldsymbol{q}_n\}$ a sequence in $C(K)$ with $\|\boldsymbol{q}_n\| \leq \hat{q}$ for all n and for some fixed $\hat{q} > 0$. If $(K_n, \boldsymbol{q}_n)$ is equi-continuous, then there is a subsequence n_k and $\boldsymbol{q} \in C(K)$ with $(K_{n_k}, \boldsymbol{q}_{n_k}) \to (K, \boldsymbol{q})$.*

References

1. Arrow KJ (2005) Personal reflections on applied general equilibrium models. In: Kehoe T, Srinivasan TN, Whalley J (eds) Frontiers in applied general equilibrium modeling. Cambridge University Press, Cambridge
2. Arrow KJ, Hahn FH (1971) General competitive analysis. Holden-Day, San Francisco
3. Anderson RM (1991) Non-standard analysis with applications to economics. In: Hildenbrand W, Sonnenschein H (eds) Handbook of mathematical economics, vol 4. North-Holland, New York, pp 2145–2208
4. Anderson RM (1992) The core in perfectly competitive economies. In: Aumann RJ, Hart S (eds) Handbook of game theory, vol 1. North-Holland, New York, pp 413–457
5. Aumann RJ (1964) Markets with a continuum of traders. Econometrica 32:39–50
6. Aumann RJ (1966) Existence of competitive equilibria in markets with a continuum of traders. Econometrica 34:1–17
7. Becker GS (1965) A theory of the allocation of time. Econ J 75:493–517
8. Becker GS (1981) A treatise on the family, Enlarged edn. Harvard University Press, Cambridge
9. Bewley TF (1972) Existence of equilibria in economies with infinitely many commodities. J Econ Theory 4:514–540
10. Bewley TF (1991) A very weak theorem on the existence of equilibria in atomless economies with infinitely many commodities. In: Khan MA, Yannelis N (eds) Equilibrium theory in infinite dimensional spaces. Springer, Berlin/New York, pp 74–101
11. Brown DJ, Robinson A (1972) A limit theorem on the cores of large standard exchange economies. Proc Nat Acad Sci USA 69:1258–1260; A Correction in 69:3068
12. Carmona G, Podczeck K (2009) On the existence of pure-strategy equilibria in large games. J Econ Theory 144:1300–1319

13. Castaneda MA, Marton J (2008) A model of commodity differentiation with indivisibilities and production. Econ Theory 34:85–106
14. Chamberlin EH (1933) The theory of monopolistic competition, 8th edn. (1965) Harvard University Press, Cambridge
15. Debreu, G. (1959) Theory of value. Wiley, New York
16. Diestel J, Uhl JJ Jr (1977) Vector measures. Mathematical surveys and monographs, vol 15. American Mathematical Society, Providence
17. Dubey P, Mas-Colell A, Shubik M (1980) Efficiency properties of strategic market games: an axiomatic approach. J Econ Theory 22:339–362
18. Forges F, Heifetz A, Minelli E (2001) Incentive compatible core and competitive equilibria in differential information economies. Econ Theory 18:349–365
19. Fradera I (1986) Perfect competition with product differentiation. Int Econ Rev 27:529–538
20. Gorman WM (1956) The demand for related goods: a possible procedure for analysing quality differentials in the egg market. Journal Paper No. 2319, Iowa Agricultural Experiment Station, Nov 1956; Reprinted in Rev Econ Stud 47:843–856 (1980)
21. Hart OD (1979) Monopolistic competition in a large economy with differentiated commodities. Rev Econ Stud 46:1–30
22. Hart S, Hildenbrand W, Kohlberg E (1974) On equilibrium allocations as distributions on the commodity space. J Math Econ 1:159–166
23. Hart S, Kohlberg E (1974) On equally distributed correspondences. J Math Econ 1:167–174
24. Hervés-Beloso C, Moreno-García E, Páscoa MR (1999) Manipulation-proof equilibrium in atomless economies with commodity differentiation. Econ Theory 14:545–563
25. Hicks JR (1956) A revision of demand theory. Clarendon Press, Oxford
26. Hildenbrand W (1974) Core and equilibria of a large economy. Princeton University Press, Princeton
27. Hotelling H (1929) Stability in competition. Econ J 39:41–57
28. Houthakker H (1952) Compensated changes in quantities and qualities consumed. Rev Econ Stud 19:155–164
29. Jones L (1983) Existence of equilibria with infinitely many consumers and infinitely many commodities. J Math Econ 12:119–138
30. Jones L (1984) A competitive model of commodity differentiation. Econometrica 52:507–530
31. Keisler HJ, Sun YN (2009) Why saturated probability spaces are necessary. Adv Math 221:1584–1607
32. Khan MA (1985) On the integration of set-valued mappings in a non-reflexive Banach space. Simon Stevin 59:257–267
33. Khan MA (1989) On Cournot-Nash equilibrium distributions for games with a nonmetrizable action space and upper semicontinuous payoffs. Trans Am Math Soc 315:127–146
34. Khan MA (2008) Perfect competition. In: Durlauf S, Blume L (eds) The new Palgrave. Macmillan, London/New York
35. Khan MA (2012) La Concorrenza Perfetta come Teoria D'ellequilibrio, Chapter 25. In: Bartocci C, Odifreddie P (eds) La Matematica, vol 4. Guilio Einaudi Editore, Roma, pp 875–947
36. Khan MA, Rath KP, Sun YN, Yu H (2015) Strategic uncertainty and the ex-post Nash property in large games. Theor Econ 10:103–129
37. Khan MA, Rath KP, Yu H, Zhang Y (2013) Large distributional games with traits. Econ Lett 118:502–505
38. Khan MA, Rath KP, Yu H, Zhang Y (2015) On the equivalence of large individualized and distributionalized games. Johns Hopkins University, mimeo
39. Khan MA, Sagara N, Suzuki T (2014) On the core and equilibria of a saturated exchange economy with differentiated commodities. The Johns Hopkins University, mimeo
40. Khan MA, Schlee E (2015) On Lionel McKenzie's 1957 intrusion into 20th-century demand theory. Canad J Econ (2015, forthcoming)

41. Khan MA, Sun YN (1987) On a symmetric Cournot-Nash equilibrium distributions in a finite-action, atomless game. B.E.B.R. Faculty working paper no. 1327, University of Illinois, Feb 1987. Also in Khan MA, Yannelis NC (eds) (1991) Equilibrium theory in infinite-dimensional spaces. Springer Verlag, New York, pp 325–333
42. Khan MA, Sun YN (1994) On large games with finite actions: a synthetic treatment. Mita J Econ 87:73–84 (in Japanese) (English translation, In: Maruyama T, Takahashi W (eds) Nonlinear and convex analysis in economic theory. Lecture notes in economics and mathematical systems, vol 419. Springer, pp 149–161)
43. Khan MA, Sun YN (1995) Extremal structures and symmetric equilibria with countable actions. J Math Econ 24:239–248
44. Khan MA, Sun YN (2002) Non-cooperative games with many players. Handbook of game theory with economic applications, vol 3. Elsevier Science, Amsterdam, pp 1761–1808
45. Khan MA, Yamazaki A (1981) On the cores of economies with indivisible commodities and a continuum of traders. J Econ Theory 24:218–225
46. Kolm S (2010) History of public economics: the historical French school. Eur J Hist Econ Thought 17:687–718
47. Lancaster K (1966) A new approach to consumer theory. J Polit Econ 74:132–157
48. Loeb PA (1975) Conversion from nonstandard to standard measure spaces and applications in probability theory. Trans Am Math Soc 211:113–122
49. Martin-da-Rocha VF (2004) Equilibrium in large economies with differentiated commodities and non-ordered preferences. Econ Theory 23:529–552
50. Mas-Colell A (1975) A model of equilibrium with differentiated commodities. J Math Econ 2:263–296
51. Mas-Colell A (1977) Indivisible commodities and general equilibrium theory. J Econ Theory 16:443–456
52. Mas-Colell A (1982) Cournotian foundations of Walrasian equilibrium theory: an exposition of recent theory, Chapter 6. In: Hildenbrand W (ed) Advances in economic theory. Cambridge University Press, Cambridge
53. Mas-Colell A (1984a) On a theorem of Schmeidler. J Math Econ 13:201–206
54. Mas-Colell A (1984b) On the theory of perfect competition. Nancy Schwartz Memorial Lecture. Kellogg School, Northwestern University
55. Mas-Colell A, Vives X (1993) Implementation in economies with a continuum of agents. Rev Econ Stud 60:613–629
56. McKenzie LW (1956–1957) Demand theory without utility index. Rev Econ Stud 24:185–189
57. McKenzie LW (2002) Classical general equilibrium theory. MIT, Cambridge
58. Milgrom P, Weber R (1985) Distributional strategies for games with incomplete information. Math Oper Res 10:619–632
59. Nikaido H (1968) Convex structures and economic theory. Academic, New York
60. Ostroy JM, Zame W (1994) Nonatomic economies and the boundaries of perfect competition. Econometrica 62:593–633
61. Podczeck K (1992) General equilibrium with differentiated commodities: the linear activity model without joint production. Econ Theory 2:247–263
62. Podczeck K (1998) Quasi-equilibrium and equilibrium in a large production economy with differentiated commodities. In: Abramovich Y, Yannelis NC, Avgerinos E (eds) Functional analysis and economic theory. Springer, Berlin/Heidelberg/New York
63. Podczeck K (2008) On the convexity and compactness of the integral of a Banach space valued correspondence. J Math Econ 44:836–852
64. Podczeck K (2010) On existence of rich Fubini extensions. Econ Theory 45:1–22
65. Pollak RA (2012) Allocating time: individuals' technologies, household technology, perfect substitutes, and specialization. Ann Econ Stat 105/106:75–97
66. Prescott E, Townsend R (1984a) Pareto optima and competitive equilibria with adverse selection and moral hazard. Econometrica 52:21–45
67. Prescott E, Townsend R (1984b) General competitive analysis in an economy with private information. Int Econ Rev 25:1–20

68. Radner R, Rosenthal R (1982) Private information and pure-strategy equilibria. Math Oper Res 7:401–409
69. Robinson J (1933) The economics of imperfect competition. Macmillan, London
70. Rosen S (1974) Hedonic prices and implicit markets: product differentiation in pure competition. J Polit Econ 82:34–55
71. Royden HW (1988) Real analysis, 3rd edn. Macmillan, London
72. Samuelson P (1969) The monopolistic competition revolution. In: Kuenne R (ed) Monopolistic competition theory: studies in impact. Wiley, New York
73. Schmeidler D (1973) Equilibrium points of non-atomic games. J Stat Phys 7:295–300
74. Sun YN, Zhang YC (2009) Individual risk and Lebesgue extension without aggregate uncertainty. J Econ Theory 144:432–443
75. Suzuki T (2000) Monopolistically competitive equilibria with differentiated commodities. Econ Theory 16:259–275
76. Suzuki T (2013a) Core and competitive equilibria for a coalitional exchange economy with infinite time horizon. J Math Econ 49:234–244
77. Suzuki T (2013b) Competitive equilibria of a large exchange economy on the commodity space ℓ^∞. Adv Math Econ 17:1–19
78. Suzuki T (2014) A coalitional production economy with infinitely many indivisible commodities. Econ Theory Bull (in press). doi:10.1007/s40505-015-0067-7
79. Theil H (1952) Qualities, prices and budget enquiries. Rev Econ Stud 19:129–147
80. Varadarajan VS (1965) Measure on topological spaces. Mathematicheskii Sbornik 55:35–100 (in Russian) (English translation, Am Math Soc Trans Ser 2 48:161–228). Reprinted in The selected works of Varadarajan VS (1988) American Mathematical Society, Providence
81. Vick JW (1994) Homology theory. Springer, Berlin/New York
82. Vind K (1964) Edgeworth allocations in an exchange economy with many traders. Int Econ Rev 5:165–177
83. Yamazaki A (1978) An equilibrium existence theorem without convexity assumptions. Econometrica 46:541–555
84. Yamazaki A (1981) Diversified consumption characteristics and conditionally dispersed endowment distribution: regularizing effect and existence of equilibria. Econometrica 49:639–654

Part III
Mini Courses

Adv. Math. Econ. 20, 131–150 (2016)

Advances in
MATHEMATICAL
ECONOMICS

Survey of the Theory of Extremal Problems

V. Tikhomirov

Abstract In the paper some general principles of the theory of extremum are considered, and basing of these principles we give a survey of fundamental results on the foundation of the theory, conditions of extrema and existence of solutions.

Keywords Theory of extremal problems • Implicit function • Convex analysis • Lagrange principle • Fields theory • Existence of solution

Article Type: Mini Course
Received: November 4, 2015
Revised: December 7 2015

Introduction

There are at least three reasons which stimulated us to solve or investigate extremal problems: pragmatical goals, desire to explain phenomenons of Nature and elementary curiosity of human race which leads them to reach the very essence of everything.

JEL Classification: C-29

Mathematics Subject Classification (2010): 26B10, 26B25, 34A55, 49-02, 49J15, 49K15, 90C25

V. Tikhomirov (✉)
Mech-math department of Lomonosov Moscow State University, 119234, Vorobyevi Gory 1, Moscow, Russia
e-mail: vmtikh@googlemail.com

S. Kusuoka, T. Maruyama (eds.), *Advances in Mathematical Economics Volume 20*, Advances in Mathematical Economics,
DOI 10.1007/978-981-10-0476-6_6

The theory of extremal problems is designed to create methods and principles for solution or investigation of concrete extremal problems. In this text the theory and its applications will be considered from general positions as a unique branch of mathematics.

Such attitude was realized in books [17] and [2] written by me with my friends and colleagues.

We will deal with the following main classes of extremal problems:

- Smooth problems with equality and inequality constraints
- Problems of calculus of variations
- Convex problems
- Problems of optimal control

In the paper the following questions will be considered:

1. Base of the theory
2. Necessary conditions
3. Sufficient conditions
4. Theory of existence
5. Algorithms
6. Application of the theory

For the extremal problem *find extremum* (i.e. maximum or minimum) *of a function* $f : X \to \mathbb{R}$ *with respect to* $x \in C$ *where* C *is a constraint*, we use the following notation:

$$f(x) \to \text{extr}, \ x \in C.$$

1 Base of the Theory

Base of the theory of extremal problems consists of *functional analysis, differential calculus* and *convex analysis*, thus the main notions of the theory are *Banach and locally convex vector spaces, differentiability* and *separability*.

Let $(X, \|\cdot\|_X)$ and $(Y, \|\cdot\|_X)$ be Banach spaces [3], V be a neighborhood of $\hat{x}$ (we denote it $V \in \mathcal{N}(\hat{x}, X)$) $F : V \to Y$ a mapping.

Definitions of differentiability. The mapping F is said to be *differentiable at a point* $\hat{x}$ ($\Leftrightarrow f \in D^1(\hat{x})$), if there exists a linear continuous operator A from X to Y ($\Leftrightarrow A \in \mathcal{L}(X, Y)$), with the following property: for any $\varepsilon > 0$ there exists $\delta > 0$ such that from inequality $\|x - \hat{x}\|_X < \delta$ it follows that $\|f(x) - F(\hat{x}) - A(x - \hat{x})\|_Y < \varepsilon \|x - \hat{x}\|_X$. The operator A is defined uniquely, it is called *derivative of mapping* F *at the point* $\hat{x}$ and is denoted $F'(\hat{x})$ [13].

The mapping F is said to be *strictly differentiable at a point* $\hat{x}$ ($\Leftrightarrow f \in SD^1(\hat{x})$) if it is differentiable at $\hat{x}$ and for any $\varepsilon > 0$ there exists $\delta > 0$ such that from inequalities $\|x_i - \hat{x}\|_X < \delta$, $i = 1, 2$ it follows that $\|f(x_2) - F(x_1) - F'(\hat{x})(x_2 - x_1)\|_Y < \varepsilon \|x_2 - x_1\|_X$ [27].

Minkowski [34] was the founder of convex analysis. The main notion of convex analysis is the notion of separability.

Definitions of separability. Let A and B be subsets of locally convex X. It is said that A *is separated from* B if there exists an element x^* from the conjugate space X^* such that $\min_{x\in A}\langle x^*, x\rangle \geq \max_{\xi\in B}\langle x^*, \xi\rangle$. In other words if there exists a hyperplane Γ such that A belongs to one of two half-spaces on which Γ divides the space X and B belongs to the other.

Main base theorems

Lemma on right-inverse mapping in linear case. *Let* $(X, \|\cdot\|_X)$ *and* $(Y, \|\cdot\|_Y)$ *be Banach spaces,* $A \in \mathcal{L}(X, Y)$. *Then there exists an operator* $R : Y \to X$ *(right-inverse) and a constant* $\gamma > 0$ *such that* $ARy = yu\|Ry\|_X \leq \gamma\| y\|_Y\ \forall y \in Y$.

Proof. From Banach open-mapping principle there exists a number $\delta > 0$, such that $AU_X(0, 1) \supset U_Y(0, \delta)$ ($U_Z(z_0, r) = \{z \in Z \mid \|z - z_0\|_Z < r\}$ is the open ball in a normed space Z with center z_0 and radius r). This means that for an arbitrary $y \in U_Y(0, \delta)$ there exists an element $x(y) \in U_X(0, 1)$, such that $Ax(y) = y$. Putting $R(y) = \frac{2\|y\|_Y}{\delta}x(\frac{\delta y}{2\|y\|_Y})$ we gain to the end. □

Right-inverse mapping theorem. *Let* $(X, \|\cdot\|_X)$ *and* $(Y, \|\cdot\|_Y)$ *be Banach spaces,* $V \in \mathcal{N}(\hat{x}, X)$, $F : V \to Y, F \in SD^1(\hat{x})$, *and moreover* $F'(\hat{x})X = Y$. *Then there exist* $\delta > 0$ *and* $K > 0$ *such that for every* $y \in Y$: $\| y-F(\hat{x})\|_Y < \delta$ *there exists an element* $x(y) \in X$ *for which* $F(x(y)) = y$ *and* $\|x(y) - \hat{x}\|_X \leq K\|f(\hat{x}) - y\|_Y$.

Prove. Let R be a right-inverse operator for $F'(\hat{x})$ from lemma. We take $\varepsilon > 0$ such that (a) $\{x \mid \|x - \hat{x}\|_X < \varepsilon\} \subset V$; (b)$\varepsilon\gamma < 1/2$, where γ is the constant from lemma. For this ε it is possible to take $\delta < \frac{\varepsilon}{8\gamma}$ such that if $\|x' - \hat{x}\|_X < \delta,\ \|x - \hat{x}\|_X < \delta$ then

$$\|f(x') - F(x) - F'(\hat{x})(x' - x)\|_Y < \varepsilon\|x' - x\|_X. \qquad (i)$$

Consider the following iterative procedure (it is called *the modify Newton method*):

$$x_{n+1} = x_n + R(y - F(x_n)),\ n \geq 0,\ x_0 = \hat{x}. \qquad (ii)$$

Then $\|x_1 - \hat{x}\|_X \leq \gamma\| y - F(\hat{x})\|_Y < \frac{\varepsilon}{8} \Rightarrow x_1 \in V$. Let $x_k \in V,\ 1 \leq k \leq n$, and we obtain

$$x_k - x_{k-1} - F'(\hat{x})(x_k - x_{k-1}) \overset{(ii)}{=} 0,\ 1 \leq k \leq n \qquad (iii)$$

and consequently

$\|x_{n+1}-x_n\|_X \overset{\text{def} R}{\leq} \gamma\| y-F(x_n)\|_Y \overset{(ii)}{=} \gamma\| y-F(x_{n)}-y+F(x_{n-1})+F'(\hat{x})(x_n-x_{n-1})\|_Y \overset{(i)}{\leq} 1/2\|x_n - x_{n-1}\|_X \leq 1/4\|x_{n-1} - x_{n-2}\|_X \leq \ldots \leq 1/2^n\|x_1 - \hat{x}\|_X$. (iv) From (iv) it follows that $\|x_n - \hat{x}\|_X = \|x_n - x_{n-1} + x_{n-1} - x_{n-2} + \ldots + x_2 - x_1 + x_1 - \hat{x}\|_X \leq (1 + 1/2 + \ldots + 1/2^n)\|x_1 - \hat{x}\|_X \leq 2\|x_1 - \hat{x}\|_X \leq \frac{\varepsilon}{4} < \varepsilon$. Hence, $\|x_j - \hat{x}\|_X <$

$\varepsilon\ \forall j \in \mathbb{N}$. Then $\|x_{n+l} - x_n|_X \le (1 + 1/2 + \ldots + 1/2^{l-1})\|x_{n+1} - x_n\|_X$, hence, $\{x_j\}_{j\in\mathbb{N}}$ is a fundamental sequence. The limit of x_n is the desired element $x(y)$. From the inequality $\|x_n - \hat{x}\|_X \le 2C\| y - F(\hat{x})\|_Y$ we obtain that $\|x(y) - \hat{x}\|_Y \le K\| y - F(\hat{x})\|_Y$, where $K = 2C$. □

Analogously it is possible to prove the following generalization of right-inverse mapping theorem.

Generalized implicit function theorem [2, 14]. *Let $\mathcal{U}$ be a topological space, X and Y Banach spaces, V a neighborhood of the point $(\hat{x}, \hat{u})$ in $X \times \mathcal{U}$, $\Psi : V \to Y$, $\Psi(\hat{x}, \hat{u}) = 0$, $\Lambda : X \to Y$ is a linear, continuous, surjective operator. Then if the mapping $u \mapsto \Psi(\hat{x}, u)$ is continuous and for any $\varepsilon > 0$ there exist a number $\delta > 0$ and a neighborhood W of the point $\hat{u}$, such that from the condition $u \in W$ and inequality $\|x_i - \hat{x}\|_Y \le \delta$, $i = 1, 2$ it follows inequality*

$$\|\Psi(x_1, u) - \Psi(x_2, u) - \Lambda(x_1 - x_2)\|_Y \le \varepsilon\|x_1 - x_2\|_X, \tag{1}$$

then for $y \in B_Y(\Psi(\hat{x}, \hat{u}), \delta_0)$, $\delta_0 = \frac{\delta(1-\theta)}{\gamma}$ the sequence of elements on X:

$$x_0 = \hat{x},\ x_n = x_{n-1} + R(y - \Psi(x_{n-1}.u)), \quad n \in \mathbb{N}, \tag{2}$$

(where γ and R are from lemma) converges to the element $\varphi(y, u) \in X$ such that $\Psi(\varphi(y, u), u) = y$ and $\|\varphi(y, u) - \hat{x}\|_X \le K\|\Psi(\hat{x}, u) - y\|_Y$, where $K = \frac{\gamma}{(1-\theta)}$.

Corollary 1 (on inverse function). *Let X and Y be Banach spaces, $V \in \mathcal{N}(\hat{x}, X)$ and $F : V \to Y$ is a continuously differentiable in V mapping with invertible derivative at $\hat{x}$. Then the mapping F is locally invertible and inverse operator $y \mapsto \varphi(y)$ is continuously differentiable in a neighborhood of a point $F(\hat{x}) \in Y$ and $\varphi'(\hat{y}) = (f'(\varphi(\hat{y})))^{-1}$.* (The first theorem of such type is due to Dini [7]).

Corollary 2 ((Lyusternik theorem on tangent space) [32] (see also [8])). *Let $(X, \|\cdot\|_X)$ and $(Y, \|\cdot\|_Y)$ be Banach spaces, $V \in \mathcal{N}(\hat{x}, X)$ and $F : V \to Y$, $F(\hat{x}) = 0$ is a mapping strictly differentiable at $\hat{x}$ with surjective differential ($F'(\hat{x})X = Y$). Then the tangent space $T_{\hat{x}}(M) = \{x \in X \mid \exists\varepsilon > 0\&\rho : [-\varepsilon, \varepsilon] \to X, \rho(t) = o(t) : \hat{x} + tx + \rho(t) \in M\}$ of zero-set $M = \{x \in X \mid F(x) = 0\}$ at $\hat{x}$ exists and is equal to the kernel of operator $F'(\hat{x})$. ($T_{\hat{x}}(M) = \mathrm{Ker}F'(\hat{x})$).*

Proof.

(a) Let $x \in T_{\hat{x}}(M)$. Then from definition of differentiability of F at $\hat{x}$, it follows that $0 = F(\hat{x} + tx + r(t)) = tF'(\hat{x})x + o(t)$, hence $x \in \mathrm{Ker}F'(\hat{x})$.
(b) Let $x \in \mathrm{Ker}F'(\hat{x})$. Consider the mapping $\Psi(x, u) = F(x + u)$. From strict differentiability of F the mapping Ψ satisfies conditions of the general implicit function theorem, therefore in accordance with this theorem there exists a mapping $\varphi : V \to X$, such that $\Psi(\varphi(x), x) = 0 \Rightarrow F(tx + \varphi(tx)) = 0 \Rightarrow \|\varphi(tx)\|_X \le K\|\Psi(tx, 0)\|_Y = K\|f(tx)\|_Y = o(t)$. □

The first separation theorem. *Let X be a locally convex space and AuB be convex subsets of X. Assume that* intA *(interior of A) is non-empty and* intA $\cap B = \emptyset$. *Then these sets can be separated. This means that there exists a non-zero element* $x^* \in X^*$ *such that*

$$\inf_{x \in A} \langle x^*, x\rangle \geq \sup_{x \in B} \langle x^*, x\rangle$$

Corollary 1 (The second separation theorem). *Convex closed set A can be strongly separated from a point* $a \notin A$. This means that there exists a non-zero element $x^* \in X^*$ such that

$$\inf_{x \in A} \langle x^*, x\rangle > \langle x^*, a\rangle$$

Proof. It is sufficient to take a neighborhood V of a such that $A \cap V = \emptyset$ and separate A and V, applying the first separation theorem. □

Corollary 2 (lemma on nontriviality of annihilator). *Let X a locally convex set and L a closed proper subspace of X* ($L \neq X$). *Then annihilator of L contains a non-zero element.*

Proof. By the data there exists an element $\hat{x} \in X \setminus L$. A functional which strongly separated L from $\hat{x}$ is a desired functional. □

2 Necessary Conditions (Principle of Lagrange)

The theory of extremum problems has its definite beginning. It happened in 1638 when Fermat (in his letter to Decartes) [6] for the problem without constraints

$$f_0(x) \to \min . \qquad (P_0)$$

de facto formulated a necessary condition for a local extremum (locextr) of problem (P_0) at a point $\hat{x}$. It is the stationary condition

$$f_0'(\hat{x}) = 0. \qquad (0)$$

This result is connected with names of Kepler [22], Fermat [6], Newton [39], Leibniz [22], Lagrange [25], Weierstrass [42], and Fréchet [13].

Kepler, who was one of the greatest calculators of all times, in [22], expressed *the calculating sense* of this formula. He wrote: "Near a maximum the decrements on both sides are in the beginning only imperceptible".

Newton [39] expressed this idea *in natural philosophy language*: "When a quantity is greatest or least at that moment its flows neither increases nor decreases".

The geometrical sense of Fermat's theorem was explained by Leibniz in his first work devoted to analysis [29]. From his considerations it follows that at an extremal point of a smooth function the tangent line is horizontal.

Fermat illustrated his method solving the following geometrical problem:

FIND A RIGHT TRIANGLE OF MAXIMUM SQUARE IF SUM OF SIDES ABOUT RIGHT ANGLE IS GIVEN.

The formalization of this problem is the following: $f_0(x) = x(a-x)$, $x \in [0, a]$. The same formalization has the unique geometrical extremal problem from Euclidian «Elements»: *In a given triangle* ABC *inscribe a parallelogram* ADEF (EF∥ AB, DE∥ AC) *of maximal area* [1].

Consider another problem without constraints which had appeared more than four centuries after the problem (P_0):

$$\int_{t_0}^{t_1} L(t, u(t))dt \to \min, \ L : [t_0, t_1] \times U \to \mathbb{R} \ (U \subset \mathbb{R}^n), \qquad (P_0')$$

where L is a continuous function. This is a criterium of minimum in (P_0'): *a piecewise-continuous function attains minimum in the problem iff the following minimum condition holds:*

$$L(t, u) \geq L(t, \hat{u}(t)) \ \forall u \in U. \qquad (0')$$

At the end of seventeenth century the theory of extremal problems made a sudden jump from one to infinity, namely from functions of one variable to functionals depending on unknown function.

It happened in 1696 when in the first scientific journal «Acta Eruditorum» John Bernoulli [4] posed the following problem:

GIVEN TWO POINTS A AND B IN A VERTICAL PLANE. FIND A PATH AMB DOWN WHICH A MOVABLE POINT M MUST BY VIRTUE OF ITS WEIGHT FALL FROM A TO B IN THE SHORTEST POSSIBLE TIME.

This challenge was called the *brachistochrone problem.*

Solution of some problems similar to brachistichrone had leaded to foundation of a special branch of mathematics which was called *calculus of variations*. The initial works in which problems analogous to brachistichrone were considered, there were memoirs of Euler (1728–1744). He deduced the necessary condition for so called *simplest problem of the calculus of variations*. This is the problem $J(x(\cdot)) = \int_{t_0}^{t_1} L(t, x(t), \dot{x}(t))dt \to \min, x(t_i) = x_i$, $i = 0, 1$. The condition for solution $\hat{x}(\cdot)$ is called now *Euler equation*: $-\frac{d}{dt}L_{\dot{x}}(t, \hat{x}(t), \dot{\hat{x}}(t)) + L_x(t, \hat{x}(t), \dot{\hat{x}}(t)) = 0$.

The investigations of Euler (added together in [11]) were continued in eighteenth century by Lagrange and Legendre and in nineteenth century by Gauss, Poisson, Jacobi, Hamilton, Poincaré, Hilbert and many others.

The next result will play an important role in our considerations:

Necessary conditions for Bolza problem (problem of calculus of variations without constraints). *Let functions* $L : \mathbb{R} \times \mathbb{R}^n \times \mathbb{R}^{n*} \to \mathbb{R}$ *and* $l : \mathbb{R}^n \times \mathbb{R}^n \to \mathbb{R}$ *be continuously differentiable in a neighborhood of the curve* $\{(t, x, y) \mid t \in \Delta = [t_0, t_1], x = \hat{x}(t), y = \dot{\hat{x}}(t)\}$, *where* $\hat{x}(\cdot) \in C^1(\Delta, \mathbb{R}^n)$ *and this function attains a local minimum (in the space* $C^1(\Delta, \mathbb{R}^n)$*) in the Bolza problem*

$$\mathcal{B}(x(\cdot)) = \int_{t_0}^{t_1} L(t, x(t), \dot{x}(t))\, dt + l(x(t_0), x(t_1)) \to \min, \tag{P_1}$$

then Euler equation

$$-\frac{d}{dt}\widehat{L}_{\dot{x}}(t) + \widehat{L}_x(t) = 0,\ \widehat{L}_{\dot{x}}(t) = L_{\dot{x}}(t, \hat{x}(t), \dot{\hat{x}}(t)),\ \widehat{L}_x(t) = L_x(t, \hat{x}(t), \dot{\hat{x}}(t)) \tag{1_a}$$

and transversality conditions

$$\widehat{L}_{\dot{x}}(t_i) = (-1)^i \hat{l}_{x(t_i)},\ i = 0, 1, \tag{1_b}$$

hold.

Method of investigation of smooth problem with equality constraints

$$f_0(x) \to \min;\ f_j(x) = 0,\ 1 \le j \le m, \tag{P_2}$$

was given by Lagrange. In his memoir [26] he wrote

> One can state the following general principle. If one is looking for the maximum or minimum of some function of many variables subject to the condition that these variables are related by a constraint given by one or more equations, then one should add to the function whose extremum is sought, functions that yield the constraint equations each multiplied by undetermined multipliers and seek the maximum or minimum as if the variables are independent. The resulting equations, combined with the constraint equations, will serve to determine all unknowns.

Let us apply this Lagrange principle to the problem (P_2).

The function $\mathcal{L}(x, \overline{\lambda}) = \sum_{i=0}^m \lambda_i f_i(x)$ is called *Lagrange function* of (P_2), vector $\overline{\lambda} = (\lambda_0, \dots, \lambda_m)$ is called *vector of Lagrange multipliers*. To minimize this Lagrange function means to solve the problem:

$$\mathcal{L}(x, \overline{\lambda}) = \sum_{i=0}^{m} \lambda_i f_i(x) \to \min.$$

It is a smooth problem without constraints, so from (0) it follows that

$$\hat{x} \in \text{locmin}(p_1) \Rightarrow \mathcal{L}_x(\hat{x}, \overline{\lambda}) = 0 \quad \Leftrightarrow \quad \sum_{i=0}^{m} \lambda_i f_i'(\hat{x}) = 0. \tag{2}$$

The equality (2) is called *Lagrange multiplier rule.*

In the middle of twentieth century were considered two new classes of problem: *convex problems* and *problems of optimal control.*

The first convex problem with economical motivation was posed by Gaspard Monge [35]. It was a special type of transportation problem:

HOW ORGANIZE THE SUPPLYING OF NECESSARY PRODUCTS TO THE POINTS OF DESTINATION FOR MAXIMUM ECONOMY?

The discrete transportation problem can be formalized like that

$$\sum_{i=1}^{N} \sum_{j=1}^{M} c_{ij} x_{ij} \to \min, \sum_{i=1}^{N} x_{ij} \le a_j, \sum_{j=1}^{M} x_{ij} \le b_i, x_{ij} \ge 0.$$

The most popular subclass of convex problems was called problems *of linear programming*:

$$f_0(x) = \sum_{i=1}^{n} c_i x_i \to \min, f_j(x) = \sum_{i=1}^{n} a_{ij} x_i \le b_i, \ 1 \le j \le m, \ x \in \mathbb{R}_+^n. \tag{P_3}$$

This type of problems which was opened by Kantorovich [20], played a fundamental role in mathematical economics.

In this case necessary conditions for solution $\hat{x}$ of the problem (P_3) correspond to the Lagrange idea in strong form, namely *there exists vector* $\overline{\lambda} = (1, \lambda_0, \ldots, \lambda_m), \lambda_0 = 1, \lambda_i \ge 0, \lambda_j f_j(\hat{x}) = 0, 1 \le j \le m$ *such that Lagrange function* $x \mapsto \mathcal{L}(x, \overline{\lambda})$ attains its minimum at $\hat{x}$:

$$\min_{x \in \mathbb{R}_+^n} \mathcal{L}(x, \overline{\lambda}) = \mathcal{L}(\hat{x}, \overline{\lambda}). \tag{3}$$

The first problem of optimal control was posed by Isaak Newton. In his great memoir [38] posed and solved the following problem of technical motivation:

FIND THE SOLID OF RESOLUTION OF GIVEN LENGTH AND WIDTH THAT IS SUBJECT TO LEAST RESISTANCE WILE MOVING IN SOME «RARE» MEDIUM.

This problem is called *Newton's aerodynamical problem.* It has the following formalization:

$$\int_0^{x_1} \frac{x dx}{1 + y'^2(x)} \to \min, \quad y(0) = 0, \ y(x_1) = y_1, \ y'(x) \ge 0.$$

Really it was the first problem of *optimal control.*

One of the main subclass of optimal control problems is formalized as follows:

$$\int_{t_0}^{t_1} f(t, x(t), u(t))dt \to \min, \ \dot{x} = \varphi(t, x, u(t)), x(t_i) = x_i, \ i = 0, 1, u(t) \in U, \tag{P_4}$$

where $x \in \mathbb{R}^n$, $u \in \mathbb{R}^r$, U is a subset of $\mathbb{R}^r$.

Such problems were introduced by Pontryagin and his followers (see [40]) in the middle of twentieth century.

The Lagrange function of (P_4) has the following standard form:

$$\mathcal{L}((x(\cdot), u(\cdot)), \overline{\lambda}) = \int_{t_0}^{t_1} (\lambda_0 L(t, x(t), \dot{x}(t), u(t)) + p(t)(\dot{x}(t) - \varphi(t, x(t), u(t))dt.$$

The Lagrange idea in this case form the following generalized vector $(\lambda_0, p(\cdot))$, where $\lambda \in \mathbb{R}, p(\cdot)$ is an n-dimensional vector-function.

To apply the Lagrange idea to the problem (P_4) it is necessary to consider two problems:

(a) $\mathcal{L}((x(\cdot), \hat{u}(\cdot)), \overline{\lambda}) \to \min,$

(b) $\mathcal{L}((\hat{x}(\cdot), u(\cdot)), \overline{\lambda}) \to \min, \quad u \in U$ or

(b') $\int_{t_0}^{t_1} (\lambda_0 L(t, \hat{x}(t), u(t)) - p(t)\varphi(t, \hat{x}(t), u(t))dt \to \min, \quad u(t) \in U.$

The problem (a) is the simplest problem of calculus of variations. Necessary condition of minimum in the simplest problem of calculus of variations according to (2) consists of the Euler equation

$$-\dot{p} - p\hat{\varphi}_x + \lambda_0 \widehat{L}_x = 0. \tag{4_a}$$

Necessary and sufficient condition in the problem (b') according to $(0')$ is the following *minimum condition:* $\lambda_0 L(t, \hat{x}(t), u) - p(t)\varphi(t, \widehat{x}(t), u) \geq \lambda_0 \widehat{L}(t) - p(t)\widehat{\varphi}(t) \quad \forall u \in U$, or in equivalent form

$$\max_{u \in U}(p(t)\varphi(t, \hat{x}(t), u) - \lambda_0 L(t, \hat{x}(t), u)) = p(t)\hat{\varphi}(t) - \lambda_0 \widehat{L}(t). \tag{4_b}$$

where $\widehat{L}(t) = L(t, \hat{x}(t), \dot{\hat{x}}(t))$ and so on. Necessary conditions (4_a) and (4_b) are called *Pontryagin maximum principle.*

The first fundamental principle:

Necessary conditions of extremum for extremal problems where smoothness is interlaced with convexity, from the times of Fermat and Lagrange till our times (till Pontryagin and his followers), correspond to general principle which we call

the Lagrange's principle of eliminating of constraints, according to which the following **Main thesis** holds:

A NECESSARY CONDITION FOR MINIMUM IN THE PROBLEM WITH CONSTRAINTS COINCIDES WITH THE NECESSARY CONDITION FOR MINIMUM OF THE LAGRANGE FUNCTION «AS IF THE VARIABLES ARE INDEPENDENT» (we quote Lagrange).

In other words (roughly speaking) *there exist Lagrange multipliers such that necessary conditions for the problem with constraints coincide with the necessary condition for minimization of Lagrange function as the problem without constraints.*

The second fundamental principle:

> Integration with respect to continuous measure leads to convexity or "almost" convexity of images of the integral mappings.

In $\mathbb{R}^n$ this idea is based on the following famous Lyapounov theorem [31]:

IMAGE OF FINITE DIMENSIONAL CONTINUOUS VECTOR-MEASURE IS A CLOSED COMPACT.

Construction of mix of control-functions involves «almost convexity» in calculus of variations and optimal control.

Lagrange Principle for smooth-convex problems

Consider the problem:

$$f_0(x) \to \min; \quad F(x, u) = 0, \qquad (\mathcal{P})$$

where X and Y be normed spaces, $\mathcal{U}$ be a set, $V \in \mathcal{O}(\hat{x}, X)$ (a neighborhood of $\hat{x}$ in X), $f_0 : V \to \mathbb{R}$, $F : V \times \mathcal{U} \to Y$ with a natural definition of a local minimum of a pair $(\hat{x}, \hat{u})$.

The Lagrange function of the problem $(\mathcal{P})$ has the natural form

$$\mathcal{L}((x, u), \bar{\lambda}) = \lambda_0 f_0(x) + \langle \lambda, F(x, u) \rangle,$$

where $\bar{\lambda} = (\lambda_0, \lambda) \in \mathbb{R}_+ \times Y^*$ (Y^* is the conjugate space), $\bar{\lambda}$ is called the *collection of Lagrange multipliers*. The following result holds:

Theorem 1 ((on smooth-convex Lagrange principle) [17]). *Let in the problem $(\mathcal{P})$ X and Y be Banach spaces with the following conditions of f and F:*

- *(a) of smoothness:* $f_0 \in D^1(\hat{x})$, $x \mapsto F(x, u) \in SD^1(\hat{x})$ $\forall u \in \mathcal{U}$;
- *(b) of weak regularity:* $\mathrm{codim} F_x(\hat{x})X < \infty$;
- *(c) of convexity are satisfied:* $F(x, \mathcal{U}) \in \mathrm{Co}(Y)$.

Then if the pair $(\hat{x}, \hat{u})$ is a local minimum in the problem $(\mathcal{P})$ then there exists a collection of Lagrange multipliers $\bar{\lambda} = (\lambda_0, \lambda) \neq 0$, $\lambda_0 \geq 0$, such that necessary conditions for this pair correspond to the Lagrange principle of the problem $(\mathcal{P})$.

This means that the following relations holds:

$$\mathcal{L}_x((\hat{x},\hat{u}),\overline{\lambda}) = 0 \Leftrightarrow \lambda_0 f_0'(\hat{x}) + (f_x(\hat{x},\hat{u}))^*\lambda = 0, \qquad (I)$$

$$\min_u \mathcal{L}((\hat{x},u),\overline{\lambda}) = \mathcal{L}((\hat{x},\hat{u}),\overline{\lambda}) \Leftrightarrow \min_u \langle \lambda, F(\hat{x},u)\rangle = 0. \qquad (II)$$

They are called a *stationary condition* and *a condition of minimum* respectively.

All results from (0) to (4), i.e. «from Fermat till Pontryagin», are corollaries of this theorem (we will show it).

Proof. Notations: $\Lambda = F_x(\hat{x},\hat{u})$, $A = F(\hat{x},\mathcal{U})$, $\Lambda X = Z$ (codimension of Z let be m). Consider factor-space X/Z. Image of the canonical mapping $\pi : Y \to X/Z$ is the space isomorphic to $\mathbb{R}^m$.

There are two possibilities: 1. The set πA doesn't contain zero as an internal point of $\mathbb{R}^m$ (degenerate case) or 2. Zero-point ia an internal point of set πA (non degenerate case).

In the first case separate zero of the space Z from the set πA by a functional $\tilde{\lambda}$ and putting $\lambda = \pi^*\tilde{\lambda}$, we obtain that for Lagrange function $\mathcal{L}(x,u,0,\lambda) = \langle \lambda, F(x,u)\rangle$ stationary condition and minimum condition were satisfied:

$\pi\Lambda X = 0 \Rightarrow \forall x \in X\ 0 = \langle\widetilde{\lambda,\pi\Lambda x}\rangle = \langle\pi^*\widetilde{\lambda},\Lambda x\rangle = \langle\lambda,\Lambda x\rangle \Rightarrow \mathcal{L}_x(\hat{x},\hat{u},0,\lambda) = 0$;

$0 \le \langle\tilde{\lambda},\pi A\rangle = \langle\pi^*\widetilde{\lambda,y}\rangle\ \forall y \in A, \Rightarrow \min_{u\in\mathcal{U}} \mathcal{L}(\hat{x},u,0,\lambda) \ge \mathcal{L}(\hat{x},\hat{u},0,\lambda)$.

Consider the second case. Let us consider the main (nondegenerate) case, when there exist $m \in \mathbb{N}$, $\{z_j\}_{j=1}^m$, $z_j \in \pi F(\hat{x},v_j)$ such that the conic hull of $\{z_j\}_{j=1}^m$ is Z. This means that there exists a vector $\overline{\xi} \neq 0$ and constants $\overline{\alpha}_i > 0$, such that $\Lambda\overline{\xi} + \sum_{i=1}^m \overline{\alpha}_i F(\hat{x},v_i) = 0$ (i).

Let v_0 be an element of $\mathcal{U}$ such that $F(\hat{x},v_0) \in Y_1$. Then there exists $x_{v_0} \in X$ for which $\Lambda x_{v_0} + F(\hat{x},v_0) = 0$. (iv). Define the mapping $\Phi : V \times \mathbb{R} \times \mathbb{R}^m \to Y$ by the formula: $\Phi(x,\alpha) = (1 - \sum_{j=0}^m \alpha_j)F(\hat{x}+x,\hat{u}) + \sum_{j=0}^m \alpha_j F(\hat{x}+x,v_j)$,

From the condition (a) of the theorem it follows that $\Phi \in SD^1((0,0),Y)$ and $\Phi'((0,0))[(x,\alpha)] = \Lambda x + \sum_{j=0}^m \alpha_j F(\hat{x},v_j)$. It is easy to prove, that $\Phi'((0,0))(X,\mathbb{R}_+^{m+1}) = Y$. So all conditions of the theorem on inverse mapping are satisfied. From (i) it follows that $\Phi'(0,0)[\hat{x} + \varepsilon\overline{\xi}, 1, \varepsilon\overline{\alpha}_1,\dots,\varepsilon\overline{\alpha}_n] = \Lambda(x_{v_0} + \varepsilon\overline{\xi}) + F(\hat{x},v_0) + \varepsilon(\sum_{i=1}^m \overline{\alpha}_i F(\hat{x},v_i)) = 0$.

From the theorem on inverse function it follows that for small t there exist $r(t)$ and $\rho_i(t)$ such that $\Phi(t(x_v + \varepsilon\overline{\xi}) + r(t), t + \rho_0(t), t\overline{\alpha}_1 + \rho_1(t),\dots,t\overline{\alpha}_m + \rho_m(t)) = 0$ and $\|r(t)\|_X + \sum_{i=0}^m |\rho_i(t)| \le K\Phi(t(x_v + \varepsilon\overline{\xi},t),\varepsilon t\overline{\alpha}_1,\dots,\varepsilon t\overline{\alpha}_m)$, hence $r(t)$ and $\rho_i(t)$ is $o(t)$. Thus (from the condition c) about the convexity or condition c') on "almost convexity" (see definition in [43])) for some $u(v,t,\beta) \in \mathcal{U}$ the equality $F(\hat{x} + tx_v + \varepsilon\overline{\xi} + r(t), u_v(t)) = 0$ holds true. We construct an admissible element $(\hat{x} + tx_v + r(t), u_v(t))$ in the problem (P), such that if $t \to 0$ then $x(t) = \hat{x} + tx_v + \varepsilon\overline{\xi} + r(t)$ tends to $\hat{x}$. We suppose that $(\hat{x},\hat{u}) \in$ *locmin* of (P), consequently $\langle f_0'(\hat{x}), x_v\rangle \ge 0$. The following implication $\Lambda x_v + F(\hat{x},v) = 0 \Rightarrow \langle f_0'(\hat{x}), x_v\rangle \ge 0$ is proved. From lemma on a kernel of a regular operator we obtain that $f_0'(x) + \Lambda^*\lambda_1^* = 0$ (v) for some $\lambda_1^* \in Y_1^*$ with the property $\langle f_0'(\hat{x}), x\rangle \geq 0$ $\forall x$ such that $\Lambda x \in F(\hat{x}, \mathcal{U})$. Application of a separation theorem leads to the main result.

□

Corollaries of Theorem 1.

1. Lagrange problem of calculus of variations

Corollary 1 (necessary conditions of minimum for Lagrange problem of calculus of variations). *Let in the problem*

$$f_0(x(\cdot), u(\cdot)) \to \min,$$

$$f_i(x(\cdot), u(\cdot)) \leq 0,\ 1 \leq i \leq m',\ f_i(x(\cdot), u(\cdot)) = 0,\ m'+1 \leq i \leq m,\ \mathcal{F}(x(\cdot), u(\cdot)) = 0 \qquad (P_1')$$

where $\Xi = C^1([t_0, t_1], \mathbb{R}^n) \times C([t_0, t_1], \mathbb{R}^r)$, $\xi = (x(\cdot), u(\cdot))$, $Y = C([t_0, t_1], \mathbb{R}^n)$ $f_i(x(\cdot), u(\cdot)) = \int\limits_{t_0}^{t_1} L_i(t, x(t), u(t))dt + l_i(x(t_0), x(t_1))$, $\mathcal{F}(x(\cdot), u(\cdot)) = \dot{x}(t) - \varphi(t, x(t), u(t))$, L_i, l_i *and* φ *are smooth enough.*

Necessary conditions for a local minimum of a pair $(\hat{x}(\cdot), \hat{u}(\cdot)) \in \Xi$ *in the problem* (P_1')*, (which is called Lagrange problem of calculus of variations) correspond to the Lagrange principle.* Namely, there exists a nonzero vector $\overline{\lambda} = (\lambda_0, \ldots, \lambda_m, p(\cdot))$ $(\lambda_0 \geq 0)$ of Lagrange multipliers, such that for Lagrange function $\mathcal{L} = \int\limits_{t_0}^{t_1} L(t, x(t), u(t))dt + l(x(t_0), x(t_1))$, where $L(t, x, u) = \sum\limits_{i=0}^{m} \lambda_i L_i(t, x, u) + \langle p(t), x(t) - \varphi(t, x(t), u(t))\rangle$, $l = \sum_{i=0}^{m} \lambda_i l_i(x(t_0), x(t_1))$, the stationary conditions for the problem (P_1') consist of Euler equation:

$$-\frac{d}{dt}\widehat{L}_{\dot{x}}(t) + \widehat{L}_x(t) = 0, \qquad (1')$$

a *transversality conditions*: $\widehat{L}_{\dot{x}}(t_i) = (-1)^i \hat{\ell}_{x(t_i)}$, $i = 0, 1$, negative condition $\lambda_i \geq 0$ and conditions of complementary slackness $\lambda_i f_i(\hat{x}(\cdot), \hat{u}(\cdot)) = 0$, $1 \leq i \leq m'$ are satisfied.

2. Smooth problems with equality and inequality constraints

Corollary 2 (Lagrange principle for smooth problems with equality and inequality constraints). *Let in the problem*

$$f_0(x) \to \min,\ f_i(x) \leq 0,\ 1 \leq i \leq m,\ \mathcal{F}(x) = 0, \qquad (P_2')$$

X and Y be Banach spaces, $f_0 \in D^1(\hat{x})$, $x \mapsto \mathcal{F}(x) \in SD^1(\hat{x}, Y)$ *and* $\operatorname{codim}\mathcal{F}_x(\hat{x})$ $X < \infty$.

Necessary condition for an element $\hat{x}$ to be a local minimum in the problem (P_2') corresponds to the Lagrange principle of the problem (P_2').

Namely, if $\hat{x}$ is a local minimum of the problem (P_2'), then for the Lagrange function $\mathcal{L}(x, \overline{\lambda}) = \sum_{i=0}^{m} \lambda_i f_i(x) + \langle \lambda, \mathcal{F}(x) \rangle$ the stationary condition

$$\mathcal{L}_x(\hat{x}, \overline{\lambda}) = 0 \Leftrightarrow \sum_{i=0}^{m} \lambda_i f_i'(\hat{x}) + (\mathcal{F}_x(\hat{x}))^* \lambda = 0, \tag{2'}$$

and conditions of *nonnegativity and complementary slackness* ($\lambda_i \geq 0,\ \lambda_i f_i(\hat{x}) = 0,\ 1 \leq i \leq m'$)) holds.

For proof of the Corollary 2 after the following reformulation of the problem (P_2') $f_0(x) \to \min;\quad F(x, u) = 0,\quad u \in \mathbb{R}_+^m$, where $F(x, u) = (f_1(x) + u_1, \ldots, f_m(x) + u_m, \mathcal{F}(x))$ application of the Theorem 1 leads to the Corollary 2.

The problem (P_1) is a smooth problem with equality and inequality constraints, so the Corollary 1 follows from Corollary 2 and necessary conditions of minimum in Bolza problem.

Historical remarks. Finite dimensional case without inequalities—Lagrange [26], infinite dimensional case without inequalities—Lyusternik [32]. Necessary conditions for smooth problems with equality and inequality constraints in finite dimensional case are due to Karush [21] and John [19], in infinite dimensional this result is due to Dubovitsky and Milyutin [10].

Almost all classical necessary conditions for weak minimum in the calculus of variations (Euler [11], Lagrange (from 1759) [25], Gauss and Ostrogradskii (30-th of 19-th century), Poisson, Weierstrass [42] and many others (see [5, 23]) are implications of Corollary 1.

3. Convex problems

Corollary 3 (criteria of minimum in convex problems). *Let in the problem*

$$f_0(u) \to \min, f_i(u) \leq 0,\ 1 \leq i \leq m',\ f_i(u) = 0,\ m' + 1 \leq i \leq m,\ u \in U, \tag{P_3'}$$

$\mathcal{U}$ be a vector space, f_i, $i = 1, \ldots, m'$ convex functions, f_i, $i = m' + 1, \ldots, m$ affine functions, on $\mathcal{U}$ and U convex subset $\mathcal{U}$.

Then if $\hat{u}$ is a solution of the problem (P_3') (i.e. it attains absolute minimum of the problem), then necessary conditions for an element $\hat{x}$ of the problem (P_3') correspond to the strong Lagrange principle in the problem (P_3').

Namely, there exists a nonlinear vector $\overline{\lambda} = (\lambda_0, \ldots, \lambda_m) \in \mathbb{R}^{m+1}$ ($\lambda_j \geq 0$) of Lagrange multipliers such that for Lagrange function $\mathcal{L}(\hat{u}, \overline{\lambda}) = \sum_{i=0}^{m} \lambda_i f_i(u)$ the minimum condition

$$\min_{u \in U} \mathcal{L}(u, \overline{\lambda}) = \mathcal{L}(\hat{u}, \overline{\lambda}) \tag{3'}$$

and conditions of complementary slackness $\lambda_i f_i(\hat{u}) = 0,\ 1 \le i \le m'$ and nonnegativity ($\lambda_0 \ge 0, \lambda_i \ge 0,\ 1 \le i \le m'$) are satisfied. (If $\lambda_0 \ne 0$ these conditions are sufficient conditions).

4. Optimal control problem

The next problem we call *optimal control problem in Pontryagin's form:*

$$\int_{t_0}^{t_1} L(t, x(t), u(t))dt \to \min,\ \dot{x} = \varphi(t, x, u(t)),\ x(t_i) = x_i, u \in U, \qquad (P_4')$$

where phase coordinates $x(t)$ and control function $u(t)$ are connected by differential relation $\dot{x} = \varphi$ resolved relatively derivative of $x(\cdot)$.

Problems of such types were starting points for problems of Optimal Control. Now we have to decipher the Lagrange principle for such problems.

The Lagrange function of (P_4') has the following standard form:

$\mathcal{L}((x(\cdot), u(\cdot)), \overline{\lambda}) = \int_{t_0}^{t_1} (\lambda_0 L(t, x(t), \dot{x}(t), u(t) + p(t)(\dot{x}(t) - \varphi(t, x(t), u(t)))dt +$ $\lambda_1 x(t_0) + \lambda_2 x(t_1)$.

The Lagrange multipliers in this case form the following generalized vector $(\lambda_0, \lambda_1, \lambda_2, p(\cdot))$, where $\lambda \in \mathbb{R}, \lambda_i,\ i = 1, 2 \in \mathbb{R}^n, p(\cdot)$ is an n-dimensional vector-function.

To apply the Lagrange principle to the problem (P_4') it is necessary to consider two problems:

(a) $\mathcal{L}((x(\cdot), \hat{u}(\cdot)), \overline{\lambda}) \to \min,$

(b) $\mathcal{L}((\hat{x}(\cdot), u(\cdot)), \overline{\lambda}) \to \min, \quad u \in U$ or

(b') $\int_{t_0}^{t_1} \lambda_0 L(t, \hat{x}(t), u(t)) - p(t)\varphi(t, \hat{x}(t), u(t)))dt \to \min, \quad u(t) \in U.$

The problem (a) is a problem of Bolza. Necessary conditions of minimum in Bolza problem consist of the Euler equation

$$-\dot{p} - p\hat{\varphi}_x + \lambda_0 \widehat{L}_x = 0. \qquad (4_a')$$

Necessary and sufficient condition in the problem (b') is the following *minimum condition:*

$$\lambda_0 L(t, \hat{x}(t), u) - p(t)\varphi(t, \hat{x}(t), u) \ge \lambda_0 \widehat{L}(t) - p(t)\hat{\varphi}(t) \quad \forall u \in U. \qquad (4_b')$$

We will prove that the following result holds:

Necessary conditions for problem of optimal control in Pontyagin's form. *Let integrand L and right part of differential relation φ are smooth enough. Then if pair $(\hat{x}(\cdot), \hat{u}(\cdot))$ furnished a strong local minimum of problem (P_4') then necessary* *conditions of this minimum corresponds with the Lagrange principle.* In other words equalities ($4'_a$) and ($4'_b$) hold.

Relations ($4'_a$) and ($4'_b$) are nothing else but so called *Pontryagin maximum principle—PMP* (Pontryagin, Boltyanski, Gamkrelidze, Mishchenko, 1961) [40].

3 Stability of Solutions and Sufficient Conditions (Ideas of Hamilton and Jacobi)

It is reasonable to include the extremal to be investigated to some family of extremals.

W. Hamilton[15]

3.1 Construction of Fields

Main thesis: NON GENERATED SECOND DERIVATIVE OF LAGRANGE FUNCTION GIVES POSSIBILITY TO CONSTRUCT FIELDS OF EXTREMALS AND CONSEQUENTLY TO PROVE STABILITY OF SOLUTION. CONSTRUCTION OF FIELDS LEADS FOR PROBLEMS OF CALCULUS OF VARIATIONS TO HAMILTON–JACOBI EQUATIONS AND TO SUFFICIENT CONDITIONS FOR EXTREMALS.

Let X and Y be Hilbert spaces, V a neighborhood of a point $\hat{x} \in X, f_0 : V \to \mathbb{R}$, $F : X \to Y$. and let f_0 and mapping $x \mapsto F(x)$ be twice differentiable in V. Consider the following problem with equalities:

$$f(x) \to \min, \ F(x) = 0. \qquad (\tilde{P})$$

The expression $\mathcal{L}(x, \lambda) = f(x) + \langle \lambda, F(x) \rangle$ is Lagrange function of problem $(\tilde{P})$ (with $\lambda_0 = 1$). Consider «a standard perturbation» of problem $(\tilde{P})$:

$$f(x) \to \min, \ F(x) = y. \qquad ((\tilde{P})_y)$$

Theorem 3.1 ((on field of extremals) [17]). *Let F be twice continuously differential in V, $F_x(\hat{x}, \hat{u})X = Y$, there exists $\hat{\lambda}$ such that $\mathcal{L}_x(\hat{x}, \hat{\lambda}) = 0$ and moreover $\mathcal{L}_{xx}(\hat{x}, \hat{\lambda})[x, x] \geq \alpha \|x\|^2 \ \forall x \in \operatorname{Ker}F'(\hat{x}) (\alpha > 0)$. Then there exists a neighborhood $W \subset Y$ of $0 \in Y$ and mapping $y \mapsto (x(y), \lambda(y))$, defined on W, such that $\mathcal{L}_x(x(y), \lambda(y)) = 0$, $x(0) = \hat{x}$, $\lambda(0) = \hat{\lambda} u F(x(y)) = y$.*

The family $\{x(y)\}$ is called *field of extremals*.

Proof. Consider mapping $\Phi(x, y) = (\mathcal{L}_x(\hat{x} + x, \widehat{\eta} + \eta), F(\hat{x} + x))$. It maps a neighborhood $\widetilde{V}$ of $(0, 0) \in X \times Y^* = X \times Y$ into $X^* \times Y = H$.

We have: $\Phi(0, 0) = (0, 0)$, $\Phi \in C^1$ and $\Phi'(0, 0)[x, \lambda] = (\mathcal{L}_{xx}(\hat{x}, \hat{\lambda})x + (f'(\hat{x}))^* \lambda, F'(\hat{x})x)$. It is possible to prove that $\Phi'(0, 0)$ is a homeomorfic mapping of

$X \times Y$ on itself and to put $\Phi^{-1}(0, y) = (x(y), \eta(y))$ for which $\mathcal{L}_x(\hat{x}+x(y), \widehat{\eta}+\eta(y)) = 0, F(\hat{x} + x(y)) = y.$ □

Put $S(y) = f(x(y))$. After differentiating we obtain that $S'(y) = -\eta(y)$.

3.2 *Conditions of Extrema for the Simplest Problem of Calculus of Variations*

Consider the following problem:

$$J(x(\cdot)) = \int_{t_0}^{t_1} L(t, x(t), \dot{x}(t))dt \to \min, \ x(t_0) = x_0, \ x(t_1) = x_1, \qquad (P_1)$$

where integrand $L : \mathbb{R}\times\mathbb{R}\times\mathbb{R}$ and a function $\hat{x}(\cdot)$, which is suspected to be minimum in the problem (P_1) (in the space $C^1([t_0, t_1])$, are smooth enough.

Let $y(\cdot) \in C_0^1([t_0, t_1]) = \{y(\cdot) \in C^1([t_0, t_1]) \mid y(t_0) = y(t_1) = 0\}$. After repeated differentiation we obtain $\mathcal{K}(y(\cdot)) = \frac{d^2\mathcal{L}(\hat{x}(\cdot)+\vartheta y(\cdot))}{d\theta^2}|_{\theta=0} = \int_{t_0}^{t_1}(A(t)\dot{y}(t)^2 + B(t)y^2(t))dt \geq 0$ (where $A(t) = \widehat{L}_{\dot{x}\dot{x}}(t)$, $B(t) = \widehat{L}_{xx} - \frac{d}{dt}\widehat{L}_{x\dot{x}}(t)$ are continuous). In other words function $\hat{y}(\cdot) \equiv 0$ is a solution of the problem

$$\int_{t_0}^{t_1} (A(t)\dot{y}^2(t) + B(t)y^2(t))dt \to \min \ y(\cdot) \in C_0^1([t_0, t_1]). \qquad (\tilde{P}_1)$$

Application of Lagrange's principle (⇔ Pontryagin maximum principle) to this problem leads to Euler equation

$$-\frac{d}{dt}(A(t)\dot{y}) + B(t)y = 0, \ y(t_0) = y(t_1) = 0$$

(it is called *Jacobi equation for problem* $(\tilde{P}_1)$) and Legendre condition $A(t) \geq 0$ [28]. We permit further that strong Lagrange condition $A(t) > 0 \ \forall t \in [t_0, t_1]$ is satisfied. There exists the unique solution $\overline{y}(\cdot)$ of Jacobi equation with boundary conditions $y(t_0) = 0, \ \dot{y}(t_0) = 1$. It is very easy to reduce from Pontryagin maximum principle, that if $\hat{x}(\cdot)$ locally minimize problem (P_1), then $\overline{y}(t) \geq 0 \ \forall t \in (t_0, t_1)$. This fact is called *Jacobi necessary condition of weak minimum in problem* (P_1)[18].

Now we suppose that *strong Jacobi condition* ($\bar{y}(t) > 0 0\ \forall t \in (t_0, t_1)$) is satisfied. In this case the quadratic functional $\mathcal{K}$ is represented in the following form:

$$\mathcal{K}(y(\cdot)) = \int_{t_0}^{t_1} A(t)(\dot{y}(t) + \frac{\dot{\bar{y}}(t)}{\bar{y}(t)} y(t))^2 dt,$$

thus it affords the strong minimum for $(\tilde{P}_1)$ and from the general theorem on field extremals it follows that $\{x(t, y) = \frac{y}{\bar{y}(t_1)}\bar{y}(t)\}$ is the field of extremals which covers all strip $(t_0, t_1] \times \mathbb{R}$. Closeness of integrants of functionals J and $\mathcal{K}$ leads to field of extremals of the problem $(\tilde{P}_1)$ which covers a neighborhood of the extremal $\hat{x}(\cdot)$. Formula for derivative of S-function if it apply to S-function of the simplest problem of CA we obtain to Hamilton–Jacobi equation and then to Weierstrass formula for increment of functional of the problem $(\tilde{P}_1)$. It leads to the following result of Weierstrass [42]:

Theorem 3.2. *Strong conditions of Lagrange and Jacobi together with quasi regularity of integrand L are sufficient conditions of strong local minimum of extremal* $\hat{x}(\cdot)$.

4 Existence (Principles of Compactness, Contractibility, Topology, Order and Monotonicity)

> I am sure that it would be possible to realize proofs of existence of solutions by means of some general principle of existence (similar to Dirichlet principle). And may be it will approach us to answer the following question: whether every regular problem has a solution (may be in some extended sense)
>
> D. Hilbert [16]

Main thesis: *Hilbert's «general principle of exitence» is in my opinion Weierstrass–Baire principle of compactness, according to which a lower semicontinuous function on a compact is bounded from below and attains its minimum.*

We illustrate application of this principle explaining existence theorem of solution of the simplest problem of calculus of variations:

$$J(x(\cdot)) = \int_{t_0}^{t_1} L(t, x(t), \dot{x}(t))dt \to \min, x(t_i) = x_i,\ i = 0, 1. \qquad (P_1)$$

Theorem 3 (on existence in calculus of variations). *If Lagrangian L in* (P_1) *satisfies conditions of quasi regularity (this means that functions* $y \mapsto L(t, x, y)$ *are* *convex) and growth (i.e. $L(t,x,y) \geq a\|y\|^p + b, a > 0, p > 1$), then the problem ($P_1$) has absolutely continuous solution.*

In this result growth provides compactness and regularity provides semicontinuity.

But really almost all results mentioned above are nothing else but *theorems on existence* (of inverse mappings, of a hyperplane which separates convex bodies, Lagrange multipliers, fields of extremals or solutions of problems). These «existences» are based on the following principles of existence:

1. **Weierstrass–Baire compacness principle**. This principle was formulated above.
2. **Brouwer topological principle**. *Let $F \in C(B^n, \mathbb{R}^n)$, $B^n = \{x \in \mathbb{R}^n \mid |x| \leq 1\}$ such that $|x - F(x)| < 1 - \delta, 0 < \delta < 1$. Then for all $y \in \mathbb{R}^n$, $|y| < \delta$ there exists $\hat{x}$ such that $y = F(x)$.*
3. **Baire ordered principle**. *Let (X, d) be a complete metric space and let $\{A_k\}_{k \in \mathrm{N}}$ be a family of nowhere dense sets in X. Then there exists a point $\hat{x} \in X \setminus \cup_{k \in \mathrm{N}} A_k$.*
4. **Minty–Brouder monotonicity principle**. *A strictly monotone, continuous and coercitive mapping $F : \mathbb{R}^n \to \mathbb{R}^n$ has a unique solution of the equation $F(x) = y$ for each $y \in \mathbb{R}^n$ (and it can be reached by means of some special iterative procedures)*
5. **Banach contraction mapping principle**. *Complete metric space with a contraction mapping admits a unique fixed-point.*

And now I want to give a short trip trough proofs of our main results.

Proofs of right-inverse mapping and generalized implicit function theorems are builded on combination of lemma on right-inverse mapping in linear case (which in turn is settled on Baire ordered principle) and modify Newton method (which is based on contractibility). Existence of Lagrange multipliers and construction of fields of extremals are based on theorems on inverse mappings. Existence of different objects in convex analysis is settled on monotonicity. Existence of solutions are relying on compactness. Convexity of images of optimal control mappings can be proved by combining of «almost convexity» of integral mappings and correction them with the help of Brouwer principle.

5 Algorithms (Methods of Solutions)

Main thesis. There are two the most important approaches to numerical solution of extremal problems. One is called *Direct methods*. It is based on approximation of infinite dimensional problems by finite dimensional ones. The second approach is connected with solution of equations which follows from necessary conditions.

In direct methods there are three the most essential ideas of reaching the goal: the idea of reasonable decent, ideas of using monotonicity and ideas of penalization.

Among the most well-known methods there are simplex method of Dantzig [9] for solution of linear programming problems (this method realizes an idea of reasonable descent), Levin–Newman section method [30, 37], ellppsoid method of Shor–Yudin–Nemirovsky (they use ideas of monotonicity), interior method of Karmarkar of convex optimization based on idea of penalization.

Application of the theory of extremal problems

Let us turn back to the beginning, where were formulated reasons which stimulated persons to solve and investigate extremal problems. Now I want to formulate talking points about this.

- Laws of dynamical processes (in optic, mechanics, hydro and aerodynamics an so on) are described by variational principles. So the mathematical theory of «naturphilosophy» (using the word of Newton) are expressed on the language of calculus of variations.
- Rules of control over dynamical processes (such as cosmic navigation) are given by variational principles with constraints which characterize limitations human possibilities, so this processes often can be formalized as problems of optimal control.
- Linear programming and optimal control became a mathematical basis of mathematical economics.
- The majority of problems which were solved without computers have a standard solution basing on principles of the theory of extremum.

References

1. Euclid (2007) Euclid's elements. Green Lion Press, Ann Arbor
2. Alexeev V, Tikhomirov V, Fomin S (1987) Optimal control. Plenum Publishing Corporation, New York
3. Banach S (1932) "Théorie des opérations linéaire", Warszava, Monographje Matematyczne
4. Bernoulli I (1696) Problema novum, ad cujus solutionem Matematici invitantur. Acta Eruditorum
5. Bliss GA (1963) Lectures on the calculus of variations. University of Chicago Press
6. de Fermat P (1891) Oeuvres de Fermat, vol 1. Gauthier-Villars, Paris
7. Dini U (1877/1878) Analisi infinitesimale. Lezzione dettate nella Università Pisa. Bd 2
8. Dmitruk AV, Milyutin AA, Osmolovskii NP (1980) Lyustrenik's theorem and the theory of extrema. Uspehi Mat Nauk 35(6): 11–46
9. Dantzig GB (1963) Linear Programming and Extensions. Princeton University Press, Princeton, NJ
10. Dubovitskii AY, Milyutin AA (1965) Extremum problems with constraints. Zh Vychisl Mat i Mat Fiz 5:395–453
11. Euler L (1744) Methodus inveniendi lineas curvas maximi minimive proprietate gaudentes sive solutio problematis isoperimetrici latssimo sencu accepti. Lausanne
12. Fenchel W (1953) Convex Cones, Sets, and Functions. Princeton University Department of Mathematics, Princeton, NJ
13. Frèchet V (1912) Sur la notion de differentielle. Nouvelle annale de mathematique, Ser. 4, V.XII. S. 845

14. Graves LM (1950) Some mapping theorems. Duke Math J 17:111–114
15. Hamilton WR (1835) Second essay on a general methods in dynamics. Philos Trans R Soc Pt 1:95–144
16. Hilbert D. Hazewinkel M (ed) (2001) "Hilbert problems", Encyclopedia of mathematics. Springer. ISBN:978-1-55608-010-4
17. Ioffe A, Tikhomirov V (1979) Theory of extremal problems. North-Holland, Amsterdam
18. Jacobi CGJ (1837) Zur Theorie der Variations-Rechnung und der Differential-Gleichungen. Krelle's Journall 17:68–82
19. John F Extremal problems with inequalities as subsidery conditions. In: Studies and Essays. Courant Anniverrsary Volume, 1948, pp 187–204
20. Kantorovich LV (1939) Mathematical methods of organizing and planning production. Manag Sci 6(4)(Jul., 1960):366–422
21. Karush WE (1939) Minima of functions of several variables with inequalities as side conditions, University of Chicago Press
22. Kepler I (1615) The volume of a Wine Barrel – Kepler's Nova stereometria doliorum vinariorum, Lincii (Roberto Cardil Matematicas Visuales)
23. Kneser A (1925) Lehrbuch der Variationsrechnung. Springer, Wiesbaden
24. Kuhn HW, Tucker AW (1951) Nonlinear programming. University of California Press, Berkley, pp 481–482
25. Lagrange JL (1766) Essai d'une nouvelle méthode pour determiner les maxima et les minima periales Petropolitanae, vol 10, 51–93
26. Lagrange JL (1797) Théorie des fonctions analytiques, Paris
27. Leach E (1961) A note on inverse mapping theorem. Proc AMS 12:694–697
28. Legendre AM (1786) Mémoire sur la maniere de distingue les maxima des minima dans le calcul de variations/ Histoire de l'Academie Royallle des Sciences. Paris, pub 1788. 7–37
29. Leibniz G (1684) Acta Eroditorum, L.M.S., т. V, pp 220–226
30. Levin AY On an algorithm for the minimization of convex function Sov. Math., Dokl. – 1965. – no. 6, pp 268–290
31. Lyapunov AA (1940) O vpolnye additivnykh vyektor-funktsiyakh: // Izvyestiya Akadyemii nauk SSSR. Syer. matyematichyeskaya. 4(6):465–478
32. Lyusternik LA (1934) On constrained exstrema of functionals (in Rusian). Matem. Sbornik 41(3):390–401. See also Russian Math Surv 35(6):11–51 (1980)
33. Mayer A (1886) Begr*ü*ndung der Lagrangesche Multiplikatorenmethode der Variatinsrehbung. Math Ann 26
34. Minkovski H (1910) Geometrie der Zahlen. Teubner, Leipzig
35. Monge G (1781) Memoire sur la théorie des déblais et des remblais, Paris
36. Moreau JJ (1964) Fonctionelles sus-differènciables. C R Acad Sci (Paris) 258:1128–1931
37. Newman DJ (1965) Location of the maximum on unimodal surfices. J ACM 12 № 3:395–398
38. Newton I (1999) Mathematical principles of natural philosophy. University of California Press, Berkeley
39. Newton I, Whiteside DT (1967–1982). The mathematical papers of Isaac Newton, 8 vols. Cambridge University Press, Cambridge. ISBN:0-521-07740-0
40. Pontryagin LS, Boltyanskii VG, Gamkrelidze RV, Mishchenko EF (1962) The mathematical theory of optimal processes (Russian). English translation: Interscience
41. Rockafellar RT (1997) Convex analysis. Princeton landmarks in mathematics. Princeton University Press, Princeton
42. Weierstrass K (1927) Mathematische Werke, Bd 7. Vorlesungemn uber Variationsrehtung. Akad. Verlag, Berlin–Leipzig
43. Avakov ER, Magaril-Il'yaev GG, Tikhomirov VM (2013) Lagrange's principle in extremum problems with constraints. Usp Mat Nauk 68(3):5–38

Adv. Math. Econ. 20, 151–180 (2016)

Advances in
MATHEMATICAL
ECONOMICS

Fourier Analysis of Periodic Weakly Stationary Processes: A Note on Slutsky's Observation

Toru Maruyama

Abstract The periodic behavior of a specific weakly stationary stochastic process (w.s.p.) is examined from a viewpoint of classical Fourier analysis.

(1) A w.s.p. has a spectral measure which is absolutely continuous with respect to the Lebesgue measure if and only if it is a moving average of a white noise. (2) A periodic or almost periodic w.s.p. must have a "discrete" spectral measure. Combining these two, we can conclude that any moving average of a white noise can neither be periodic nor almost periodic.

However any w.s.p. can be approximated by a sequence of almost periodic w.s.p.'s in some specific sense.

Keywords Weakly stationary process • Periodicity • Almost periodicity • Spectral measure

Article Type: Mini Course
Received: October 8, 2015
Revised: October 28, 2015

T. Maruyama (✉)
Keio University, Tokyo, Japan
e-mail: maruyama@econ.keio.ac.jp

S. Kusuoka, T. Maruyama (eds.), *Advances in Mathematical Economics Volume 20*, Advances in Mathematical Economics,
DOI 10.1007/978-981-10-0476-6_7

Introduction

During a decade around 1930, the world economy was thrown into the serious depression, nobody had ever experienced.

A number of economic theories were proposed in order to analyze and control the violence of business fluctuations. Among them, the work of Eugen Slutsky [30], an Ukrainian mathematician, is particularly remarkable from the mathematical point of view.[1] It was published in Econometrica, 1937.

He tried to explain more or less regular fluctuations of macro-economic movements based upon the overlapping effects of random shocks.

However, frankly speaking, his analysis was rather experimental and devoid of mathematical rigor.

The purpose of the present paper is to give a systematic exposition of the mathematical skeleton of the Slutsky problem being based upon classical Fourier analysis.

Our main conclusions are two-fold.

(1) Any weakly stationary stochastic process generated by moving average of white noise (call Slutsky process) can neither be periodic nor almost periodic. If a weakly stationary process is periodic or almost periodic, its spectral measure must be discrete (i.e. a weighted sum of Dirac measures). However the spectral measure of a Slutsky process must be absolutely continuous with respect to the Lebesgue measure and hence it has a spectral density function. It immediately follows that any Slutsky process can never be periodic and even the almost periodicity is impossible. The Bochner-Herglotz theory concerning Fourier representation of positive semi-definite functions provide us the key analytical tool.

(2) Slutsky's conclusion is correct in a sophisticated sense. That is, the spectral measure of any weakly stationary process can be approximated by a sequence of spectral measures of almost periodic processes.

The earlier draft of this paper was read at the 20th Conference of the International Federation of Operational Research Societies, which was held in Barcelona, (July 13–18, 2014). It is a pleasure for me to express my cordial gratitude to the late Professor Tatsuo Kawata for incessant encouragement, which led me to the field of Fourier analysis. I am much indebted to Professor Shigeo Kusuoka for his kind suggestions concerning probability theory, which I am not familiar with very well. Helpful comments by Dr. Yuhki Hosoya and Mr.Chaowen Yu are also gratefully acknowledged.

JEL Classification: CO2, E32

Mathematics Subject Classification (2010): 42A38, 42A82, 60G10

[1]Frisch [11] also deserves a special attention.

These positive claims are developed in the last two sections. I am much indebted to several forerunners in this discipline, in particular to the late Professor Tatsuo Kawata for basic ideas. I have to confess, however, that there still remain complicated reasonings which do not seem convincing enough for me. I try to reorganize these puzzling aspects as coherently as possible.

The first six sections cover the fundamental concepts and results which are not so familiar to mathematical economists but are indispensable prerequisites for understanding last two sections.[2]

1 Basic Concepts

Let $(\Omega, \mathscr{E}, P)$ be a probability space, and T a subset of $\mathbb{R}$. T is usually interpreted as the space of time and, in this paper, T is assumed to be either $\mathbb{R}$ or $\mathbb{Z}$. A function $X : T \times \Omega \to \mathbb{C}$ is said to be a stochastic process if the function $\omega \mapsto X(t, \omega)$ is $(\mathscr{E}, \mathscr{B}(\mathbb{C}))$-measurable for any fixed $t \in T$, where $\mathscr{B}(\mathbb{C})$ is the Borel σ-field in $\mathbb{C}$. $\mathbb{C}, \mathbb{R}$ and $\mathbb{Z}$ denote the sets of complex numbers, real numbers and integers, respectively.

The expectation $\mathbb{E}X(t, \omega)$ of the stochastic process $X(t, \omega)$ at $t \in T$ is defined by

$$\mathbb{E}X(t, \omega) = \int_\Omega X(t, \omega)dP.$$

The function $\rho : T \times T \to \mathbb{C}$ defined by

$$\rho(s, t) = \mathbb{E}[X(s, \omega) - \mathbb{E}X(s, \omega)]\overline{[X(t, \omega) - \mathbb{E}X(t, \omega)]}, \quad s, t \in T$$

is called the covariance function of $X(t, \omega)$.

Let $\mathscr{T}$ be the set of all the finite tuples of elements of T, that is

$$\mathscr{T} = \{\mathbf{t} = (t_1, t_2, \cdots, t_n) \mid t_j \in T, \quad j = 1, 2, \cdots, n, \quad n \in \mathbb{N}\}.$$

$X_{\mathbf{t}}(\omega)$ denotes the vector

$$X_{\mathbf{t}}(\omega) = (X(t_1, \omega), X(t_2, \omega), \cdots, X(t_n, \omega)), \quad \mathbf{t} \in \mathscr{T}.$$

The set function $\nu_{X_{\mathbf{t}}} : \mathscr{B}(\mathbb{C}^n) \to \mathbb{R}$ defined by

$$\nu_{X_{\mathbf{t}}}(E) = P\{\omega \in \Omega | X_{\mathbf{t}}(\omega) \in E\}, \quad E \in \mathscr{B}(\mathbb{C}^n)$$

[2]Kawata [17, 18], Maruyama [23] and Wold [31] are classical works on Fourier analysis of stationary stochastic processes, which provided me with all the basic mathematical background. Among more recent literatures, I wish to mention Brémaud [5]. Granger and Newbold [12] Chap. 2, Hamilton [13] Chap. 3 and Sargent [26] Chap. XI are textbooks written from the standpoint of economics.

is a measure on $\mathscr{B}(\mathbb{C}^n)$, called the distribution of $X_{\mathbf{t}}(\omega)$. The set of all the distributions $\{\nu_{X_{\mathbf{t}}} | \mathbf{t} \in \mathscr{T}\}$ is called the system of finite dimensional distributions determined by $X(t,\omega)$.

A stochastic process $X(t,\omega)$ is strongly stationary if the distribution of $(X(t_1 + t,\omega), X(t_2 + t,\omega), \cdots, X(t_n + t,\omega))$ is independent of t for any $\mathbf{t} = (t_1, t_2, \cdots, t_n) \in \mathscr{T}$.

One of the most important concepts of this paper is that of weakly stationary processes. $X(t,\omega)$ is said to be **weakly stationary** if the following conditions are satisfied.

(i) The absolute moment of the second order is finite:

$$\mathbb{E}|X(t,\omega)|^2 < \infty \quad \text{for each} \quad t \in T.$$

(ii) The expectation is constant throughout time:

$$\mathbb{E}X(t,\omega) = m(t) = m \quad \text{constant for all} \quad t \in T.$$

(iii) The covariance depends only upon the difference $u = s - t$ of times :

$$\rho(s,t) = \rho(s+h, t+h) \quad \text{for any} \quad s,t,h \in T.$$

The condition (iii) permits us to denote the covariance by $\rho(u)$ instead of $\rho(s,t)$ for the sake of simplicity.

It is well-known that the strong stationarity implies the weak stationarity provided that $\mathbb{E}|X(t_0,\omega)|^2 < \infty$ for some $t_0 \in T$ (cf. Itô [15] pp. 236–237).

We add two more concepts, the strong continuity and the measurability, in the case of $T = \mathbb{R}$. Propositions 1 and 2 clarify the relation between these concepts. (See Kawata [19] pp. 53–55 and Crum [8] for details.)

If a function $A : \mathbb{R} \to \mathfrak{L}^2(\Omega, \mathbb{C})$ defined by

$$A : t \mapsto X(t,\omega)$$

is continuous at a point $t_0 \in \mathbb{R}$; *i.e.*

$$\mathbb{E}|X(t,\omega) - X(t_0,\omega)|^2 \to 0 \quad \text{as} \quad t \to t_0,$$

we say that $X(t,\omega)$ is strongly continuous at t_0. If $X(t,\omega)$ is strongly continuous at every $t \in \mathbb{R}$, $X(t,\omega)$ is said to be strongly continuous on $\mathbb{R}$.

If $X(t,\omega)$ is $(\mathscr{L} \otimes \mathscr{E}, \mathscr{B}(\mathbb{C}))$-measurable, $X(t,\omega)$ is called a measurable stochastic process, where $\mathscr{L}$ is the Lebesgue σ-field on $\mathbb{R}$.

Proposition 1. *Let* $X : \mathbb{R} \times \Omega \to \mathbb{C}$ *be a weakly stationary process with* $\mathbb{E}X(t,\omega) = 0$ *for all* $t \in \mathbb{R}$.

(i) $X(t, \omega)$ *is strongly continuous at* $t = 0$ *if and only if* $X(t, \omega)$ *is uniformly strongly continuous on* $\mathbb{R}$.
(ii) $\rho(u)$ *is continuous at* $u = 0$ *if and only if* $\rho(u)$ *is uniformly continuous on* $\mathbb{R}$.
(iii) $X(t, \omega)$ *is strongly continuous if and only if* $\rho(u)$ *is uniformly continuous on* $\mathbb{R}$.

Proposition 2. *The covariance function of any measurable and weakly stationary process* $X(t, \omega)$ *is continuous. So* $X(t, \omega)$ *is strongly continuous on* $\mathbb{R}$.

2 Linear Stochastic Processes

One of the most typical examples of weakly stationary processes is so called the linear stochastic process (the Slutsky process, informally).

We assume, for a while, the case of discrete time; i.e. $T = \mathbb{Z}$. And, according to the ordinary custom, we write $X_n(\omega)$ rather than $X(n, \omega)$. Let $X_n : \Omega \to \mathbb{C}(n \in \mathbb{Z})$ be a stochastic process such that

(i) $\mathbb{E}X_n(\omega) = 0 \quad \text{for all} \quad n \in \mathbb{Z}$,
(ii) $\mathbb{E}|X_n(\omega)|^2 = \sigma^2$ (σ^2 is a constant, called the variance, which is independent of n), and
(iii) $\mathbb{E}X_n(\omega)\overline{X_m(\omega)} = 0 \quad \text{for all} \quad m \neq n$.

A stochastic process which satisfies (i), (ii) and (iii) above is called a white noise. The condition (iii) is called the serially uncorrelatedness.

The next proposition confirms that some average of a white noise is weakly stationary.[3]

Proposition 3. *Assume that a stochastic process* $X_n : \Omega \to \mathbb{C}(n \in \mathbb{Z})$ *satisfies (i), (ii) and (iii), and* $\{c_\nu\}_{\nu \in \mathbb{Z}}$ *is a sequence with* $\sum_{n=-\infty}^{\infty} |c_\nu|^2 < \infty$ *(i.e.* $\{c_\nu\} \in \ell_2(\mathbb{C})$*). Define another stochastic process* $Y_n(\omega)$ *by*

$$Y_n(\omega) = \sum_{k=-\infty}^{\infty} c_{k-n} X_k(\omega), \quad n \in \mathbb{Z}. \tag{1}$$

Then the right-hand side of (1) is $\mathfrak{L}^2$*-convergent and* $Y_n(\omega)(n \in \mathbb{Z})$ *is a weakly stationary process with*

$$\mathbb{E}Y_n(\omega) = 0, \quad n \in \mathbb{Z}, \tag{2}$$

[3]cf. Kawata [19] p. 46.

and

$$\rho(u) = \sigma^2 \sum_{\nu=-\infty}^{\infty} c_{u+\nu}\bar{c}_\nu. \tag{3}$$

The weakly stationary process defined by (1) is called the **moving average process** or the **linear stochastic process** induced by $X_n(\omega)$ $(n \in \mathbb{Z})$. Proposition 1 says that a moving average process induced by a white noise is weakly stationary.

The corresponding concept of linear stochastic processes in continuous-time case $(T = \mathbb{R})$ will be defined later (cf. Theorem 2 in Sect. 7) because some additional prerequisite is required.

3 Spectral Measures

Let T be equal to $\mathbb{Z}$ or $\mathbb{R}$. A function $f : T \to \mathbb{C}$ is said to be positive semi-definite if

$$\sum_{i,j=1}^{n} f(t_i - t_j)\lambda_i\bar{\lambda_j} \geqq 0$$

for any finite tuple $\mathbf{t} = (t_1, t_2, \cdots, t_n) \in \mathscr{T}$ of elements of T and $\lambda_1, \lambda_2, \cdots, \lambda_n \in \mathbb{C}$.

The celebrated theorems due to Herglotz [14] and Bochner [2, 3] tell us that any positive semi-definite function on T can be expressed as the Fourier transform of certain positive Radon measure.[4]

Herglotz Theorem. *The following two statements are equivalent for a function* $f : \mathbb{Z} \to \mathbb{C}$.

[4]Schwartz [27] Chap. VII, §9, and Lax [20] Chap. 14 discuss the Bochner-Herglotz theorem in the spirit of the theory of distributions. Rudin [25] Chap. 1 provides a proof based upon the theory of Banach algebras. Naimark [24] Chap. 6 gives an abstract version, following D.A. Raikov.

> "Let G be a locally compact commutative topological group with unit e. Then any continuous positive semi-definite function $\varphi : G \to \mathbb{C}$ is uniquely representable in the form
>
> $$\varphi(g) = \int_{\widehat{G}} \chi(g)d\mu(\chi),$$
>
> where μ is a measure on the character group $\widehat{G}$ of G, which satisfies $\mu(\widehat{G}) = \varphi(e)$."
>
> The converse also holds true.

(i) f is positive semi-definite.
(ii) There exists a positive Radon measure μ *on* $\mathbb{T} = [-\pi, \pi]$ *which satisfies*

$$f(n) = \frac{1}{\sqrt{2\pi}} \int_{\mathbb{T}} e^{-inx} d\mu(x).$$

Bochner Theorem. *The following two statements are equivalent for a bounded continuous function* $f : \mathbb{R} \to \mathbb{C}$.

(i) f is positive semi-definite.
(ii) There exists a positive Radon measure μ *on* $\mathbb{R}$ *which satisfies*

$$f(t) = \frac{1}{\sqrt{2\pi}} \int_{\mathbb{R}} e^{-itx} d\mu(x).$$

Let $X : T \times \Omega \to \mathbb{C}$ be a weakly stationary stochastic process. Then it is easy to see that its covariance function $\rho(u)$ is positive semi-definite. In fact,

$$\begin{aligned}\sum_{i,j=1}^{n} \rho(t_i - t_j)\lambda_i\overline{\lambda_j} &= \sum_{i,j=1}^{n} \mathbb{E}X(t_i, \omega)\overline{X(t_j, \omega)}\lambda_i\overline{\lambda_j} \\ &= \mathbb{E}\left|\sum_{i,j=1}^{n} X(t_j, \omega)\lambda_j\right|^2 \geqq 0\end{aligned}$$

for any $t_j \in T$ and $\lambda_j \in \mathbb{C}$ $(j = 1, 2, \cdots, n)$.

Proposition 4. *If* $X : \mathbb{Z} \times \Omega \to \mathbb{C}$ *is a weakly stationary process, its covariance function* $\rho(u)$ *can be expressed as the Fourier transform of certain positive Radon measure* ν *on* $\mathbb{T}$:

$$\rho(n) = \frac{1}{\sqrt{2\pi}} \int_{\mathbb{T}} e^{-in\theta} d\nu(\theta).$$

This is an immediate consequence of the Herglotz Theorem and the positive semi-definiteness of ρ. There remains nothing to be explained more.

However we have to check the continuity of ρ in the case $T = \mathbb{R}$.

Proposition 4'. *If* $X : \mathbb{R} \times \Omega \to \mathbb{C}$ *is a measurable and weakly stationary process, its covariance function* $\rho(u)$ *can be expressed as the Fourier transform of certain positive Radon measure* ν *on* $\mathbb{R}$:

$$\rho(u) = \frac{1}{\sqrt{2\pi}} \int_{\mathbb{R}} e^{-iut} d\nu(t).$$

Since $X(t, \omega)$ is measurable, $\rho(u)$ is continuous by Proposition 2. We already confirmed the positive semi-definiteness of $\rho(u)$. Hence Proposition 4' follows from the Bochner Theorem.

The Radon measure ν appearing Proposition 4 or Proposition 4' is called the **spectral measure** of $X(t, \omega)$. If ν is absolutely continuous with respect to the Lebesgue measure, the Radon-Nikodým derivative is called the **spectral density function** of $X(t, \omega)$. (See Sect. 7.)

It is well-known in Fourier analysis that if the Fourier transform $\hat{\mu}$ of a Radon measure μ on $\mathbb{R}$ or $\mathbb{T}$ is zero, then $\mu = 0$ (cf. Katznelson [16] p. 164). Hence the mapping $\mu \mapsto \hat{\mu}$ is injective.

We next consider the converse problem. When a positive Radon measure ν on T is given, does there exist a weakly stationary stochastic process the spectral measure of which is exactly ν? The following proposition answers this question positively.

Proposition 5. *A positive Radon measure* ν *on* $\mathbb{T}$ *is assumed to be given. Then there exist a probability space* $(\Omega, \mathscr{E}, P)$ *and a stochastic process* $X_n : \Omega \to \mathbb{C} (n \in \mathbb{Z})$ *which satisfies the following conditions.*

(i) $X_n(\omega)$ *is a weakly stationary process.*
(ii) ν *is the spectral measure of* $X_n(\omega)$.

The next proposition is a continuous-time ($T = \mathbb{R}$) version of the above.

Proposition 5'. *A positive Radon measure* ν *on* $\mathbb{R}$ *is assumed to be given. Then there exist a probability space* $(\Omega, \mathscr{E}, P)$ *and a stochastic process* $X : \mathbb{R} \times \Omega \to \mathbb{C}$ *which satisfies the following conditions.*

(i) $X(t, \omega)$ *is a measurable and weakly stationary process.*
(ii) ν *is the spectral measure of* $X(t, \omega)$.

4 Spectral Representation of Weakly Stationary Processes

We define the subclass $\mathscr{B}^*(\mathbb{R})$ of the Borel σ-field on $\mathbb{R}$ by $\mathscr{B}^*(\mathbb{R}) = \{S \in \mathscr{B}(\mathbb{R}) | m(S) < \infty\}$. $m(\cdot)$ is the Lebesgue measure. $(\Omega, \mathscr{E}, P)$ is a probability space.

A function $\xi : \mathscr{B}(\mathbb{R}) \times \Omega \to \mathbb{C}$ is called an **orthogonal measure** (or $\mathfrak{L}^2$-orthogonal measure) if the following three conditions are satisfied.

(i) The function $\omega \mapsto \xi(S, \omega)$ is in $\mathfrak{L}^2(\Omega, \mathbb{C})$ for every $S \in \mathscr{B}^*(\mathbb{R})$.

(ii) If $S_1, S_2, \cdots \in \mathscr{B}^*(\mathbb{R})$ are mutually disjoint and $\bigcup_{n=1}^{\infty} S_n \in \mathscr{B}^*(\mathbb{R})$, then

$$\xi(\bigcup_{n=1}^{\infty} S_n, \omega) = \sum_{n=1}^{\infty} \xi(S_n, \omega) \text{ in } \mathfrak{L}^2(\Omega, \mathbb{C}).$$

(iii) If $S, S' \in \mathscr{B}^*(\mathbb{R})$ and $S \cap S' = \emptyset$, then

$$\mathbb{E}\xi(S, \omega)\overline{\xi(S', \omega)} = 0.$$

If (i), (ii) and (iii) hold true on $\mathscr{B}(\mathbb{R})$ instead of $\mathscr{B}^*(\mathbb{R})$, $\xi(S, \omega)$ is called a **finite orthogonal measure**. In this case, the set function $\nu_\xi : \mathscr{B}(\mathbb{R}) \to \mathbb{R}$ defined by

$$\nu_\xi(S) = \|\xi(S, \omega)\|_2^2, \quad S \in \mathscr{B}(\mathbb{R}) \tag{1}$$

is a finite measure on $(\mathbb{R}, \mathscr{B}(\mathbb{R}))$.

In fact, it is obvious that $\nu_\xi(S) \geqq 0$ for all $S \in \mathscr{B}(\mathbb{R})$. Furthermore, if $S_n \in \mathscr{B}(\mathbb{R})$ $(n = 1, 2, \cdots)$ are mutually disjoint, then

$$\begin{aligned} \nu_\xi(\bigcup_{n=1}^{\infty} S_n) &= \left\| \xi(\bigcup_{n=1}^{\infty} S_n, \omega) \right\|_2^2 \underset{\text{(ii)}}{=} \lim_{n\to\infty} \left\| \sum_{j=1}^{n} \xi(S_j, \omega) \right\|_2^2 \\ &\underset{\text{(iii)}}{=} \lim_{n\to\infty} \sum_{j=1}^{n} \|\xi(S_j, \omega)\|_2^2 = \sum_{n=1}^{\infty} \nu_\xi(S_n). \end{aligned}$$

This prove the σ-additivity of $\nu_\xi(\cdot)$.

We define the concept of integration with respect to the finite orthogonal measure $\xi(S, \omega)$ for functions $f \in \mathfrak{L}^2_{\nu_\xi}(\mathbb{R}, \mathbb{C})$ ($\mathfrak{L}^2$-space w.r.t. ν_ξ). We start with defining the integration for simple functions, and then proceed to general $\mathfrak{L}^2_{\nu_\xi}$-functions based upon extention by continuity. See Brémaud [5] pp. 137–141, Doob [9] Chap. IX and Kawata [18] pp. 28–35 for details.

It can easily be proved that

$$\mathbb{E}\int_{\mathbb{R}} f(x)\xi(dx,\omega)\int_{\mathbb{R}}\overline{g(x)\xi(dx,\omega)} = \int_{\mathbb{R}} f(x)\overline{g(x)}d\nu_\xi(x) \quad \text{for any} \quad f,g \in \mathfrak{L}^2_{\nu_\xi}(\mathbb{R},\mathbb{C}). \tag{2}$$

In particular,

$$\left\| \int_{\mathbb{R}} f(x)\xi(dx,\omega) \right\|_2^2 = \int_{\mathbb{R}} |f(x)|^2 d\nu_\xi(x) \quad \text{for } f \in \mathfrak{L}^2_{\nu_\xi}(\mathbb{R},\mathbb{C}). \tag{3}$$

For the sake of later reference, we call (2) and (3) by the name "the Doob-Kawata formulas" (D-K formulas).

The next theorem due to Cramér [7] and Kolmogorov[5] gives a spectral representation of weakly stationary processes.[6]

Cramér-Kolmogorov Theorem. *Let $X : \mathbb{R}\times\Omega \to \mathbb{C}$ be a measurable and weakly stationary process with $\mathbb{E}X(t,\omega) = a$ (for all $t \in \mathbb{R}$). Then there exists an orthogonal measure $\xi : \mathscr{B}(\mathbb{R})\times\Omega \to \mathbb{C}$ which satisfies*

$$X(t,\omega) = a + \frac{1}{\sqrt{2\pi}}\int_{\mathbb{R}} e^{-i\lambda t}\xi(d\lambda,\omega) \tag{4}$$

and $\nu = \nu_\xi$. Conversely, the stochastic process $X(t,\omega)$ represented by the above formula (4) in terms of an orthogonal measure $\xi(S,\omega)$ is weakly stationary.

ξ is uniquely determined corresponding to X.

The same result holds true for the case $T = \mathbb{Z}$. In this case, $\mathscr{B}(\mathbb{R})$ should be replaced by $\mathscr{B}(\mathbb{T})$ and the scope of integration in (4) should be $\mathbb{T}$ instead of $\mathbb{R}$.

5 Periodicity of Weakly Stationary Processes

Let $X : T\times\Omega \to \mathbb{C}$ be a weakly stationary process with the covariance function $\rho(u)$. $X(t,\omega)$ is called a **periodic weakly stationary process** with period τ or, in short, τ-periodic if the $\rho(u)$ is periodic with period τ; i.e. $\rho(u+\tau) = \rho(u)$.

---

[5]According to Itô [15] p. 255, Kolmogorov's important article was published in C.R. Acad. Sci. URSS, **26** (1940), 115–118. However I have never read it yet, very regrettably. That is why I dropped it from the reference list.

[6]In case $X(t,\omega)$ is real-valued, is it possible to give an expression of it in terms of an orthogonal measure without complex function like $e^{i\lambda t}$? This probelm was advocated by Slutsky [28, 29] and completed by Doob [10] and Maruyama [23]. See also Itô [15] pp. 263–266.

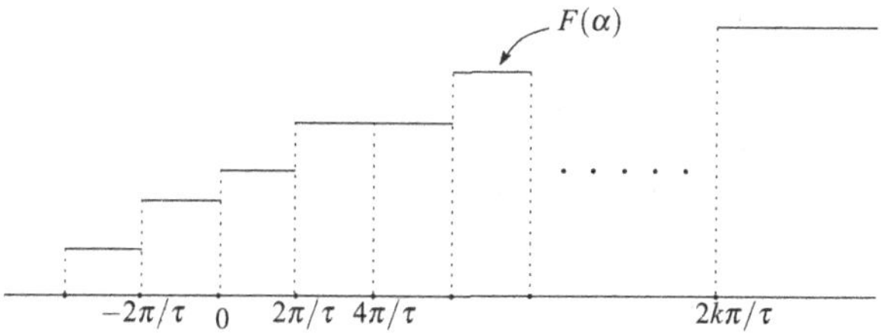

Fig. 1 Spectral density function of a periodic process

Proposition 6 is a characterization of periodic processes in discrete-time case. Proposition 6' is its continuous-time version.

Proposition 6. *Let $X : \mathbb{Z}\times\Omega \to \mathbb{C}$ be a weakly stationary process with the spectral measure ν. Then the following three statements are equivalent.*

(i) $X_n(\omega)$ *is* τ*-periodic.*
(ii) $X_{n+\tau}(\omega) - X_n(\omega) = 0 \quad a.e.(\omega) \quad$ *for all* $n \in \mathbb{Z}$.
(iii) If $E \in \mathscr{B}(\mathbb{T})$ *and* $E \cap \{2k\pi/\tau | k \in \mathbb{Z}\} = \emptyset$, *then* $\nu(E) = 0$.

Proposition 6'. *Let $X : \mathbb{R}\times\Omega \to \mathbb{C}$ be a measurable and weakly stationary process with the spectral measure ν. Then the following three statements are equivalent.*

(i) $X(t,\omega)$ *is* τ*-periodic.*
(ii) $X(t+\tau,\omega) - X(t,\omega) = 0.\ a.e.(\omega) \quad$ *for all* $\quad t \in \mathbb{R}$.
(iii) If $E \in \mathscr{B}(\mathbb{R})$ *and* $E \cap \{2k\pi/\tau | k \in \mathbb{Z}\} = \emptyset$, *then* $\nu(E) = 0$.

Remark. The spectral distribution function of $X(t,\omega)$ is defined by

$$F(\alpha) = \nu((-\infty,\alpha]), \quad \alpha \in \mathbb{R}.$$

Then (iii) means that $F(\alpha)$ is a step function with possible discontinuities at $\{2k\pi/\tau | k \in \mathbb{Z}\}$. (See Fig. 1)

If $X(t,\omega)$ is a τ-periodic weakly stationary process, the spectral measure concentrates on a countable set in $\mathbb{T}$ or $\mathbb{R}$, informally called the energy set of $X(t,\omega)$, such that the distance of any adjacent two points is some multiple of $2\pi/\tau$. We have to keep in mind that the periodic weakly stationary process can not have spectral density function.

6 Almost Periodicity of Weakly Stationary Processes

Let $f : \mathbb{R} \to \mathbb{C}$ be a function and ε a positive real number. $\tau \in \mathbb{R}$ is called an ε-**almost period** of f if

$$\sup_{x\in\mathbb{R}} |f(x-\tau) - f(x)| < \varepsilon.$$

f is said to be (uniformly) **almost periodic** in the sense of Bohr [4] if

(i) f is continuous, and
(ii) there exists $\Lambda = \Lambda(\varepsilon, f) \in \mathbb{R}$ for each $\varepsilon > 0$ such that any interval, the length of which is $\Lambda(\varepsilon, f)$, contains ε-almost period of f.

The set $\mathfrak{AP}(\mathbb{R}, \mathbb{C})$ of all the almost periodic functions of $\mathbb{R}$ into $\mathbb{C}$ forms a closed subalgebra of $\mathfrak{L}^\infty(\mathbb{R}, \mathbb{C})$.[7]

Let $X : T \times \Omega \to \mathbb{C}$ be a weakly stationary process with the covariance function $\rho(u)$. $X(t, \omega)$ is called an **almost periodic weakly stationary process** if $\rho(u)$ is almost periodic.

The following propositions play similar roles with those of Propositions 6 and 6' in Sect. 5.

Proposition 7. *Let $X : \mathbb{Z} \times \Omega \to \mathbb{C}$ be a weakly stationary process with the spectral measure ν. Then the following three statements are equivalent.*

(i) $X_n(\omega)$ is an almost periodic process.
(ii) There exists $\Gamma = \Gamma(\varepsilon, X) \in \mathbb{R}$ for each $\varepsilon > 0$ such that any interval, the length of which is $\Gamma(\varepsilon, X)$, contains some $\tau \in \mathbb{Z}$ which satisfies

$$\sup_{n \in \mathbb{Z}} \mathbb{E}|X_{n+\tau}(\omega) - X_n(\omega)|^2 < \varepsilon.$$

(iii) ν is discrete.

Proposition 7'. *Let $X : \mathbb{R} \times \Omega \to \mathbb{C}$ be a measurable and weakly stationary process with the spectral measure ν. Then the following three statements are equivalent.*

(i) $X(t, \omega)$ is an almost periodic process.
(ii) There exists $\Gamma = \Gamma(\varepsilon, X) \in \mathbb{R}$ for each $\varepsilon > 0$ such that any interval, the length of which is $\Gamma(\varepsilon, X)$, contains some $\tau \in \mathbb{R}$ which satisfies

$$\sup_{t \in \mathbb{R}} \mathbb{E}|X(t + \tau, \omega) - X(t, \omega)|^2 < \varepsilon.$$

(iii) ν is discrete.

7 Spectral Density Functions

As we saw in previous sections, *any periodic or almost periodic weakly stationary process can not have spectral density functions.*

---

[7]See Katznelson [16] p. 194. Loomis [21] is also beneficial.

Hence a problem arises: under what conditions does a weakly stationary process have a spectral density function?[8]

Let $\xi : \mathscr{B}(\mathbb{T}) \times \Omega \to \mathbb{C}$ be an orthogonal measure with $\mathbb{E}\xi(S,\omega) = 0$ for any $S \in \mathscr{B}(\mathbb{T})$. There exist a probability space $(\Omega', \mathscr{E}', P')$ and a weakly stationary process $Y_n(\omega') : \Omega' \to \mathbb{C} (n \in \mathbb{Z})$, the spectral measure of which is the Lebesgue measure m on $\mathbb{T}$ (by Proposition 5). We denote by $\mathbb{E}_\omega$ (resp. $\mathbb{E}_{\omega'}$) the expectation operator on $(\Omega, \mathscr{E}, P)$ (resp. $(\Omega', \mathscr{E}', P')$). The Cramér-Kolmogorov Theorem assures the representation

$$Y_n(\omega') = \frac{1}{\sqrt{2\pi}} \int_{\mathbb{T}} e^{-i\lambda n} \eta(d\lambda, \omega')$$

for some orthogonal measure $\eta : \mathscr{B}(\mathbb{T}) \times \Omega' \to \mathbb{C}$. (We assume that $\mathbb{E}_{\omega'} Y_n(\omega') = 0$.) η satisfies

(a) $\mathbb{E}_{\omega'} \eta(S, \omega') = 0$ for any $S \in \mathscr{B}(\mathbb{T})$, and
(b) $\mathbb{E}_{\omega'} |\eta(S, \omega')|^2 = m(S)$ for any $S \in \mathscr{B}(\mathbb{T})$.

In order to express some relations between ξ and η, we need the "adjunction" method[9] as follows. $\mathbb{E}_{(\omega,\omega')}$ denotes the expectation operator on the product probability space $(\Omega \times \Omega', \mathscr{E} \otimes \mathscr{E}', P \otimes P')$. $1(\omega)$ (resp. $1(\omega')$) denotes the constant function which is identically 1 on Ω (resp. on Ω'). Then the following conditions hold true.

(i) $\mathbb{E}_{(\omega,\omega')} \eta(S, \omega') 1(\omega) = \mathbb{E}_{\omega'} \eta(S, \omega') = 0$ for any $S \in \mathscr{B}(\mathbb{T})$.
(ii) $\mathbb{E}_{(\omega,\omega')} |\eta(S, \omega') 1(\omega)|^2 = \nu_\eta(S) = m(S)$ for any $S \in \mathscr{B}(\mathbb{T})$.
(iii) $\mathbb{E}_{(\omega,\omega')} \xi(S, \omega) 1(\omega') \cdot \overline{\eta(S', \omega') 1(\omega)} = \mathbb{E}_\omega \xi(S, \omega) \mathbb{E}_{\omega'} \overline{\eta(S', \omega')} = 0$ for any S and $S' \in \mathscr{B}(\mathbb{T})$.

Theorem 1. *Let $X_n(\omega)$ $(n \in \mathbb{Z})$ be a weakly stationary process with the spectral measure ν. Assume also that $\mathbb{E}X_n(\omega) = 0$. Thus the following two statements are equivalent.*

(i) $X_n(\omega)$ has the spectral density function.
(ii) $X_n(\omega)$ is a linear stochastic process; that is there exist a sequence $\{c_n\}$ of complex numbers such that

$$\sum_{n=-\infty}^{\infty} |c_n|^2 < \infty$$

[8]This problem was studied by Doob [9] Chap. X, §8, Chap. XI, §8 and Kawata [19] pp. 69–73. I try to clarify the subtle details embedded in their works.

[9]See Doob [9] Chap. II, §2.

and a stochastic process $Z_n(\omega)(n \in \mathbb{Z})$ *with*

$$\begin{aligned} \mathbb{E}Z_n(\omega) = 0, \quad \mathbb{E}|Z_n(\omega)|^2 = 1, \\ \mathbb{E}Z_n(\omega)\overline{Z_m(\omega)} = 0 \quad \text{if} \quad n \neq m \end{aligned} \tag{1}$$

which satisfy

$$X_n(\omega) = \sum_{k=-\infty}^{\infty} c_{k-n} Z_k(\omega) \quad a.e.$$

(The convergence on the right-hand side is in $\mathfrak{L}^2(\Omega, \mathbb{C})$*.)*

Proof. (i)⇒(ii): Let $p(\lambda) \geqq 0$ be the Radon-Nikodým derivative of ν. We also define

$$\alpha(\lambda) = \sqrt{p(\lambda)}. \tag{2}$$

Since $\alpha(\lambda) \in \mathfrak{L}^2(\mathbb{T}, \mathbb{C})$ (actually real-valued), it can be expanded by the Fourier series[10]:

$$\alpha(\lambda) = \frac{1}{\sqrt{2\pi}} \sum_{k=-\infty}^{\infty} \alpha_k e^{ik\lambda}, \tag{3}$$

$$\sum_{k=-\infty}^{\infty} |\alpha_k|^2 < \infty, \tag{4}$$

where α_k is the Fourier coefficient corresponding to $(1/\sqrt{2\pi})e^{ik\lambda} (k \in \mathbb{Z})$.

By the Cramér-Kolmogorov theorem, $X_n(\omega)$ is represented as the Fourier transform of some orthogonal measure $\xi(S, \omega)$. According to the argument preceding the theorem, there exist a probability space $(\Omega', \mathscr{E}', P')$ and an orthogonal measure $\eta(S, \omega') : \mathscr{B}(\mathbb{T}) \times \Omega' \to \mathbb{C}$ which satisfies (i), (ii) and (iii) mentioned above (p. 163).

Divide $\mathbb{T}$ into $\mathbb{T}_+$ and $\mathbb{T}_0$ defined as $\mathbb{T}_+ = \{t \in \mathbb{T} | \alpha(t) > 0\}$, $\mathbb{T}_0 = \{t \in \mathbb{T} | \alpha(t) = 0\}$.

[10]The convergence of the series in (3) is in $\mathfrak{L}^2(\mathbb{T}, \mathbb{C})$. However the series is, actually, convergent a.e. and is equal to $\alpha(\lambda)$ according to the Carleson theorem. cf. Carleson [6].

We define a couple of functions, $\alpha_1(\lambda)$ and $\alpha_2(\lambda)$, by

$$\alpha_1(\lambda) = \begin{cases} \frac{1}{\alpha(\lambda)} & \text{on } \mathbb{T}_+, \\ 0 & \text{on } \mathbb{T}_0, \end{cases} \tag{5}$$

$$\alpha_2(\lambda) = \begin{cases} 0 & \text{on } \mathbb{T}_+, \\ 1 & \text{on } \mathbb{T}_0. \end{cases}$$

Furthermore define a function $\gamma' : \mathscr{B}(\mathbb{T}) \times \Omega \times \Omega' \to \mathbb{C}$ by [11]

$$\begin{aligned} \gamma'(S, \omega, \omega') &= \int_S \alpha_1(\lambda)\xi(d\lambda, \omega)1(\omega') + \int_S \alpha_2(\lambda)\eta(d\lambda, \omega')1(\omega). \\ &= \int_S \alpha_1(\lambda)\xi(d\lambda, \omega) + \int_S \alpha_2(\lambda)\eta(d\lambda, \omega'). \end{aligned} \tag{6}$$

For the definition (6) to be possible, we have to check that $\alpha_1(\cdot) \in \mathfrak{L}^2_{\nu_\xi}$ and $\alpha_2(\cdot) \in \mathfrak{L}^2_{\nu_\eta}$ where

$$\nu_\xi(S) = \mathbb{E}_\omega|\xi(S, \omega)|^2 \quad \text{and} \quad \nu_\eta(S) = \mathbb{E}_{\omega'}|\eta(S, \omega')|^2, \quad S \in \mathscr{B}(\mathbb{T}),$$

respectively. But there is no difficulty in checking it as shown in the following:

$$\begin{aligned} \int_S |\alpha_1(\lambda)|^2 d\nu_\xi &= \int_{S\cap\mathbb{T}_+} \frac{1}{\alpha(\lambda)^2} p(\lambda) dm(\lambda) \\ &= \int_{S\cap\mathbb{T}_+} dm(\lambda) = m(S \cap \mathbb{T}_+) < \infty, \end{aligned} \tag{7}$$

$$\int_S |\alpha_2(\lambda)|^2 d\nu_\eta = \int_{S\cap\mathbb{T}_0} dm(\lambda) = m(S \cap \mathbb{T}_0) < \infty.$$

[11] In case $\mathbb{T}_0 = \emptyset$ (and so $\alpha(\lambda)$ never vanishes), the discussion becomes much easier, since it is enough to define $\gamma(S, \omega)$ simply by

$$\gamma(S, \omega) = \int_S \frac{1}{\alpha(\lambda)} \xi(d\lambda, \omega)$$

for any $S \in \mathscr{B}(\mathbb{T})$. Clearly $\nu_\gamma(S) = m(S)$.

$\gamma'(S,\omega,\omega')$ is an orthogonal measure. For instance, the orthogonality is proved as follows. Let S and $S' \in \mathscr{B}(\mathbb{T})$ be disjoint. Then we obtain

$$\begin{aligned}
&\mathbb{E}_{(\omega,\omega')}\gamma'(S,\omega,\omega')\overline{\gamma'(S',\omega,\omega')} \\
&\quad = \mathbb{E}_{\omega}\int_S \alpha_1(\lambda)\xi(d\lambda,\omega)\int_{S'}\alpha_1(\lambda)\overline{\xi(d\lambda,\omega)} \\
&\quad + \mathbb{E}_{(\omega,\omega')}\int_S \alpha_1(\lambda)\xi(d\lambda,\omega)\int_{S'}\alpha_2(\lambda)\overline{\eta(d\lambda,\omega')} \\
&\quad + \mathbb{E}_{(\omega,\omega')}\int_S \alpha_2(\lambda)\eta(d\lambda,\omega')\int_{S'}\alpha_1(\lambda)\overline{\xi(d\lambda,\omega)} \\
&\quad + \mathbb{E}_{\omega'}\int_S \alpha_2(\lambda)\eta(d\lambda,\omega')\int_{S'}\alpha_2(\lambda)\overline{\eta(d\lambda,\omega')}.
\end{aligned} \tag{8}$$

The second and third terms of (8) are zero because of the orthogonality of ξ and η in the sense of (iii) (p. 163). The first and the forth terms are also zero because $S \cap S' = \emptyset$.

By a similar computation as in (7) and (8), we have [12]

$$\mathbb{E}_{(\omega,\omega')}|\gamma'(S,\omega,\omega')|^2 = m(S), \quad S \in \mathscr{B}(\mathbb{T}). \tag{9}$$

Finally a function $\gamma : \mathscr{B}(\mathbb{T}) \times \Omega \to \mathbb{C}$ is defined by

$$\gamma(S,\omega) = \mathbb{E}_{\omega'}\gamma'(S,\omega,\omega'). \tag{10}$$

Taking account of the properties of γ', γ is shown to be an orthogonal measure with $\nu_\gamma(S) = m(S \cap \mathbb{T}_+)$.

12
$$\begin{aligned}
&\mathbb{E}_{(\omega,\omega')}|\gamma'(S,\omega,\omega')|^2 \\
&= \mathbb{E}_{\omega}\left|\int_S \alpha_1(\lambda)\xi(d\lambda,\omega)\right|^2 \\
&\quad + \mathbb{E}_{\omega'}\left|\int_S \alpha_2(\lambda)\eta(d\lambda,\omega')\right|^2 \\
&\quad + 2\mathscr{R}e\,\mathbb{E}_{(\omega,\omega')}\int_S \alpha_1(\lambda)\xi(d\lambda,\omega)1(\omega')\int_S \overline{\alpha_2(\lambda)\eta(d\lambda,\omega')1(\omega)} \\
&= \int_S |\alpha_1(\lambda)|^2\nu_\xi(d\lambda) + \int_S |\alpha_2(\lambda)|^2\nu_\eta(d\lambda) \\
&\quad + 2\mathscr{R}e\,\mathbb{E}_{(\omega,\omega')}\int_{S\cap\mathbb{T}_+}\alpha_1(\lambda)\xi(d\lambda,\omega)\int_{S\cap\mathbb{T}_0}\overline{\alpha_2(\lambda)\eta(d\lambda,\omega')} \\
&= m(S\cap\mathbb{T}_+) + m(S\cap\mathbb{T}_0) + 0.
\end{aligned}$$

The Cramér-Kolmogorov representation theorem gives [13]

$$\begin{aligned} X_n(\omega) &= \frac{1}{\sqrt{2\pi}} \int_{\mathbb{T}} e^{-in\lambda} \xi(d\lambda, \omega) \\ &= \frac{1}{\sqrt{2\pi}} \int_{\mathbb{T}} e^{-in\lambda} \alpha(\lambda) \gamma(d\lambda, \omega). \end{aligned} \tag{11}$$

Since the Fourier series (3) converges to $\alpha(\lambda)$ in $\mathfrak{L}^2$, it follows that

$$X_n(\omega) = \frac{1}{2\pi} \sum_{k=-\infty}^{\infty} \int_{\mathbb{T}} \alpha_k e^{-i(n-k)\lambda} \gamma(d\lambda, \omega) \ a.e. \tag{12}$$

In fact, (12) is verified by the computation:

$$\begin{aligned} &\mathbb{E}_\omega \left| X_n(\omega) - \frac{1}{2\pi} \sum_{k=-p}^{p} \int_{\mathbb{T}} \alpha_k e^{i(k-n)\lambda} \gamma(d\lambda, \omega) \right|^2 \\ &= \mathbb{E}_\omega \left| X_n(\omega) - \frac{1}{2\pi} \int_{\mathbb{T}} \sum_{k=-p}^{p} \alpha_k e^{i(k-n)\lambda} \gamma(d\lambda, \omega) \right|^2 \\ &= \mathbb{E}_\omega \left| \frac{1}{\sqrt{2\pi}} \int_{\mathbb{T}} e^{-in\lambda} \left[\alpha(\lambda) - \frac{1}{\sqrt{2\pi}} \sum_{k=-p}^{p} \alpha_k e^{ik\lambda} \right] \gamma(d\lambda, \omega) \right|^2 \quad \text{(by (11))} \\ &= \frac{1}{2\pi} \int_{\mathbb{T}} \left| \alpha(\lambda) - \frac{1}{\sqrt{2\pi}} \sum_{k=-p}^{p} \alpha_k e^{ik\lambda} \right|^2 d\nu_\gamma(\lambda) \qquad \text{(by D-K formula)} \end{aligned}$$

---

[13]

$$\begin{aligned} &\int_{\mathbb{T}} e^{-in\lambda} \alpha(\lambda) \gamma(d\lambda, \omega) \\ &\quad = \int_{\mathbb{T}_+} e^{-in\lambda} \alpha(\lambda) \cdot \frac{1}{\alpha(\lambda)} \xi(d\lambda, \omega) + \mathbb{E}_{\omega'} \int_{\mathbb{T}_0} e^{-in\lambda} \alpha(\lambda) \cdot \alpha_2(\lambda) \eta(d\lambda, \omega') \\ &\quad = \int_{\mathbb{T}_+} e^{-in\lambda} \xi(d\lambda, \omega) = \int_{\mathbb{T}} e^{-in\lambda} \xi(d\lambda, \omega). \end{aligned}$$

The final equality is justified by

$$\begin{aligned} \mathbb{E}_\omega \left| \int_{\mathbb{T}_0} e^{-in\lambda} \xi(d\lambda, \omega) \right|^2 &= \int_{\mathbb{T}_0} \nu_\xi(d\lambda) \quad \text{(by D-K formula)} \\ &= \int_{\mathbb{T}_0} p(\lambda) dm = 0 \quad (p(\lambda) = 0 \text{ on } \mathbb{T}_0). \end{aligned}$$

$$
\begin{aligned}
&= \frac{1}{2\pi}\int_{\mathbb{T}_+}\left|\alpha(\lambda) - \frac{1}{\sqrt{2\pi}}\sum_{k=-p}^{p}\alpha_k e^{ik\lambda}\right|^2 dm(\lambda) \\
&\leqq \frac{1}{2\pi}\int_{\mathbb{T}}\left|\alpha(\lambda) - \frac{1}{\sqrt{2\pi}}\sum_{k=-p}^{p}\alpha_k e^{ik\lambda}\right|^2 dm(\lambda) \\
&\to 0 \quad \text{as} \quad p \to \infty \qquad\qquad \text{(by (3))}.
\end{aligned}
$$

If we define

$$
Z_j(\omega) = \frac{1}{\sqrt{2\pi}}\int_{\mathbb{T}} e^{ij\lambda}\gamma(d\lambda,\omega), \tag{13}
$$

and $c_k = (1/\sqrt{2\pi})\alpha_k$, then

$$
X_n(\omega) = \sum_{k=-\infty}^{\infty} c_k Z_{k-n}(\omega) = \sum_{j=-\infty}^{\infty} c_{n+j}Z_j(\omega) \quad (\text{in } \mathfrak{L}^2). \tag{14}
$$

It is easy to check that $Z_n(\omega)$ satisfies the condition (1). $\mathbb{E}Z_j(\omega) = 0$ is clear from the definition (10) of γ ($\mathbb{E} = \mathbb{E}_\omega$). The variance$=1$ and the orthogonality come from

$$
\begin{aligned}
&2\pi\mathbb{E}Z_n(\omega)\overline{Z_m(\omega)} \\
&=\mathbb{E}\int_{\mathbb{T}} e^{in\lambda}\gamma(d\lambda,\omega)\int_{\mathbb{T}} e^{-im\lambda}\overline{\gamma(d\lambda,\omega)} \\
&= \int_{\mathbb{T}} e^{i(n-m)\lambda}d\nu_\gamma = \int_{\mathbb{T}_+} e^{i(n-m)\lambda}dm(\lambda) \\
&= \int_{\mathbb{T}} e^{i(n-m)\lambda}dm(\lambda) - \int_{\mathbb{T}_0} e^{i(n-m)\lambda}d\nu_\gamma = \int_{\mathbb{T}} e^{i(n-m)\lambda}dm(\lambda) \\
&=\delta_{n,m}\times 2\pi.
\end{aligned}
$$

The second equality is justified by D-K formula.

(ii)⇒(i): Conversely, assume that $X_n(\omega)$ is a moving average of a stochastic process $Z_n(\omega)$ which satisfies (1). Since $Z_n(\omega)$ is weakly stationary, there exists an orthogonal measure $\xi(S,\omega)$ on $\mathscr{B}(\mathbb{T})\times\Omega$ such that

$$
Z_n(\omega) = \frac{1}{\sqrt{2\pi}}\int_{\mathbb{T}} e^{-in\lambda}\xi(d\lambda,\omega) \tag{15}
$$

by the Cramér-Kolmogorov theorem. If we define $\nu_\xi(S) = \|\xi(S,\omega)\|_2^2$ as usual, ν_ξ is the spectral measure of $Z_n(\omega)$, which has the spectral density function $1/\sqrt{2\pi}$ (constant function). [14]

Consequently, we obtain, for some $\{c_n\} \in \ell_2(\mathbb{C})$, that

$$X_n(\omega) = \underset{N\to\infty}{\text{l.i.m.}} \frac{1}{\sqrt{2\pi}} \sum_{k=-N}^{N} c_{k-n} \int_{\mathbb{T}} e^{-ik\lambda} \xi(d\lambda,\omega) \tag{16}$$

$$= \underset{N\to\infty}{\text{l.i.m.}} \frac{1}{\sqrt{2\pi}} \int_{\mathbb{T}} \sum_{k=-N}^{N} c_{k-n} e^{-ik\lambda} \xi(d\lambda,\omega).$$

Since $\{c_n\} \in \ell_2$, c_n is the Fourier coefficient of some $C(\lambda) \in \mathfrak{L}^2(\mathbb{T},\mathbb{C})$, and[15]

$$\left\| \frac{1}{\sqrt{2\pi}} \sum_{k=-p}^{q} c_k e^{-ik\lambda} - C(\lambda) \right\|_2 \to 0 \quad \text{as} \quad p,q \to \infty. \tag{17}$$

Hence (writing $k - n = j$),

$$\sum_{k=-N}^{N} c_{k-n} e^{-ik\lambda} = e^{-in\lambda} \sum_{j=-N-n}^{N-n} c_j e^{-ij\lambda}$$

tends to $\sqrt{2\pi} e^{-in\lambda} C(\lambda)$ in $\mathfrak{L}^2$ as $N \to \infty$ for fixed n; i.e.

$$\left\| \sum_{k=-N}^{N} c_{k-n} e^{-ik\lambda} - \sqrt{2\pi} e^{-in\lambda} C(\lambda) \right\|_2 \to 0 \quad \text{as} \quad N \to \infty. \tag{18}$$

[14]Since $Z_n(\omega)$ is a white noise, the covariance is given by

$$\mathbb{E} Z_{n+u}(\omega)\overline{Z_n(\omega)} = \begin{cases} 1 & \text{if} \quad u = 0, \\ 0 & \text{if} \quad u \neq 0. \end{cases}$$

[15]The convergence of $(1/\sqrt{2\pi}) \sum_{k=-p}^{q} c_k e^{-ik\lambda}$ to $C(\lambda)$ also holds true "almost everywhere" thanks to the Carleson theorem. Hence c_k is the Fourier coefficient of $C(\lambda)$ corresponding to $(1/\sqrt{2\pi})e^{-ik\lambda}$.

Taking account of the fact that the density function of ν_ξ is $1/\sqrt{2\pi}$ as remarked above, we have

$$\begin{aligned}
&\mathbb{E}\left|X_n(\omega) - \int_{\mathbb{T}} C(\lambda)e^{-in\lambda}\xi(d\lambda,\omega)\right|^2 \\
=&\mathbb{E}\left|\underset{N\to\infty}{\text{l.i.m.}}\frac{1}{\sqrt{2\pi}}\int_{\mathbb{T}}\sum_{k=-N}^{N} c_{k-n}e^{-ik\lambda}\xi(d\lambda,\omega) - \int_{\mathbb{T}} C(\lambda)e^{-in\lambda}\xi(d\lambda,\omega)\right|^2 \quad \text{(by (16))} \\
=& \lim_{N\to\infty}\frac{1}{2\pi}\mathbb{E}\left|\int_{\mathbb{T}}\left[\sum_{j=-N-n}^{N-n} c_j e^{-ij\lambda} - \sqrt{2\pi}C(\lambda)\right]e^{-in\lambda}\xi(d\lambda,\omega)\right|^2 \\
=& \lim_{N\to\infty}\frac{1}{2\pi}\int_{\mathbb{T}}\left|\sum_{j=-N-n}^{N-n} c_j e^{-ij\lambda} - \sqrt{2\pi}C(\lambda)\right|^2 d\nu_\xi \quad \text{(by D-K formula)} \\
=& \lim_{N\to\infty}\int_{\mathbb{T}}\left|\frac{1}{\sqrt{2\pi}}\sum_{j=-N-n}^{N-n} c_j e^{-ij\lambda} - C(\lambda)\right|^2 \frac{1}{\sqrt{2\pi}}dm(\lambda) \\
=&0 \quad \text{(by (18))}.
\end{aligned}$$

Hence we obtain

$$X_n(\omega) = \frac{1}{\sqrt{2\pi}}\int_{\mathbb{T}}\sqrt{2\pi}C(\lambda)e^{-in\lambda}\xi(d\lambda,\omega). \quad a.e.$$

If we define $\theta(S,\omega)$ by

$$\theta(S,\omega) = \int_S \sqrt{2\pi}C(\lambda)\xi(d\lambda,\omega), \quad S \in \mathscr{B}(\mathbb{T}),$$

then $\theta(S,\omega)$ is an orthogonal measure[16] and

$$X_n(\omega) = \frac{1}{\sqrt{2\pi}}\int_{\mathbb{T}} e^{-in\lambda}\theta(d\lambda,\omega).$$

[16]The orthogonality, for instance, can be verified as follows. If S and $S' \in \mathscr{B}(\mathbb{T})$ are disjoint,

$$\begin{aligned}
\mathbb{E}\theta(S,\omega)\overline{\theta(S',\omega)} &= \mathbb{E}\int_{\mathbb{T}} C(\lambda)\chi_S(\lambda)\xi(d\lambda,\omega)\int_{\mathbb{T}}\overline{C(\lambda)\chi_{S'}(\lambda)\xi(d\lambda,\omega)} \\
&= \int_{\mathbb{T}} |C(\lambda)|^2\chi_S(\lambda)\chi_{S'}(\lambda)d\nu_\xi = 0 \quad \text{(by D-K formula)}.
\end{aligned}$$

This is the spectral representation of $X_n(\omega)$ in terms of $\theta(S,\omega)$. Consequently the spectral measure ν of $X_n(\omega)$ is given by

$$\nu(S) = \mathbb{E}|\theta(S,\omega)|^2 = 2\pi \int_S |C(\lambda)|^2 d\nu_\xi$$
$$= \sqrt{2\pi} \int_S |C(\lambda)|^2 dm(\lambda),$$

which is, of course, absolutely continuous with respect to m. **Q.E.D.**

We prepare a lemma to be used in the proof of Theorem 2. We denote by $\mathfrak{F}_2$ (resp. $\mathfrak{F}_2^{-1}$) the Fourier transform (resp. inverse Fourier transform) in the sense of Plancherel.[17]

Lemma 1.

$$\frac{1}{\sqrt{2\pi}} \int_{\mathbb{R}} f(\lambda) g(\lambda) e^{-iu\lambda} dm(\lambda) = \frac{1}{\sqrt{2\pi}} \langle \mathfrak{F}_2^{-1} f(\lambda - u), \mathfrak{F}_2^{-1} \bar{g}(\lambda) \rangle$$
$$= \frac{1}{\sqrt{2\pi}} \int_{\mathbb{R}} f(\lambda - u) g(\lambda) dm(\lambda)$$

for any f *and* $g \in \mathfrak{L}^2(\mathbb{R},\mathbb{C})$.

(We have to note $f \cdot g \in \mathfrak{L}^1(\mathbb{R},\mathbb{C})$. $\langle \cdot,\cdot \rangle$ *denotes the inner product in* $\mathfrak{L}^2(\mathbb{R},\mathbb{C})$.*)*

Proof. By definition of the inverse Fourier transform in the sense of Plancherel, we have

$$\mathfrak{F}_2^{-1}(f(z)e^{-izu})(\lambda) = \underset{A\to\infty}{\text{l.i.m.}} \frac{1}{\sqrt{2\pi}} \int_{-A}^{A} f(z) e^{-izu} e^{iz\lambda} dm(z)$$
$$= \underset{A\to\infty}{\text{l.i.m.}} \frac{1}{\sqrt{2\pi}} \int_{-A}^{A} f(z) e^{i(\lambda-u)z} dm(z) \tag{19}$$
$$= \mathfrak{F}_2^{-1} f(\lambda - u).$$

[17] We can also establish

$$\frac{1}{\sqrt{2\pi}} \int_{\mathbb{R}} f(\lambda) g(\lambda) e^{-iz\lambda} dm(\lambda) = \frac{1}{\sqrt{2\pi}} (\mathfrak{F}_2 f * \mathfrak{F}_2 g)(z).$$

Taking account of $f \cdot g \in \mathfrak{L}^1$, we obtain

$$\frac{1}{\sqrt{2\pi}} \int_{\mathbb{R}} f(z) \cdot g(z) e^{-izu} dm(z)$$

$$= \frac{1}{\sqrt{2\pi}} \langle f(z) e^{-izu}, \bar{g}(z) \rangle$$

$$= \frac{1}{\sqrt{2\pi}} \langle \mathfrak{F}_2^{-1}(f(z)e^{-izu})(\lambda), \mathfrak{F}_2^{-1}\bar{g}(\lambda) \rangle$$

(by Parseval equality)

$$= \frac{1}{\sqrt{2\pi}} \langle \mathfrak{F}_2^{-1} f(\lambda - u), \mathfrak{F}_2^{-1}\bar{g}(\lambda) \rangle \quad \text{(by (19))}$$

$$= \frac{1}{\sqrt{2\pi}} \langle f(\lambda - u), \bar{g}(\lambda) \rangle \quad \text{(by Parseval)}$$

$$= \frac{1}{\sqrt{2\pi}} \int_{\mathbb{R}} f(\lambda - u) g(\lambda) dm(\lambda).$$

Q.E.D.

Theorem 2. *Let $X(t, \omega)(t \in \mathbb{R})$ be a measurable and weakly stationary process with the spectral measure ν. Assume also that $\mathbb{E}X(t, \omega) = 0$. Then the following two statements are equivalent.*

(i) $X(t, \omega)$ has the spectral density function.
(ii) There exists a measurable and weakly stationary process $X' : \mathbb{R} \times \Omega \to \mathbb{C}$ of the form

$$X'(t, \omega) = \int_{\mathbb{R}} w(\lambda - t) \gamma(d\lambda, \omega) \ a.e. \ (\omega),$$

the spectral measure of which is ν, where $w(\cdot) \in \mathfrak{L}^2(\mathbb{R}, \mathbb{C})$ and the orthogonal measure $\gamma(S, \omega) : \mathscr{B}(\mathbb{R}) \times \Omega \to \mathbb{C}$ satisfies

$$\nu_\gamma(S) = \mathbb{E}|\gamma(S, \omega)|^2 = m(S \cap \mathbb{R}_0), \ S \in \mathscr{B}^*(\mathbb{R}).$$

(The definition of $\mathbb{R}_0$ is given below.)

Proof. By the Cramér-Kolmogorov theorem, there exists an orthogonal measure $\xi(S,\omega)$ on $\mathscr{B}(\mathbb{R})\times\Omega$ such that

$$X(t,\omega)=\frac{1}{\sqrt{2\pi}}\int_{\mathbb{R}} e^{-i\lambda t}\xi(d\lambda,\omega).$$

(i)⇒(ii): Let $p(\lambda)\geqq 0$ be the Radon-Nikodým derivative of ν. If we define

$$\alpha(\lambda)=\sqrt{p(\lambda)},$$

then $\alpha\in\mathfrak{L}^2(\mathbb{R},\mathbb{R})$.

The covariance $\rho(u)$ of $X(t,\omega)$ can be written as

$$\begin{aligned}\rho(u)&=\frac{1}{\sqrt{2\pi}}\int_{\mathbb{R}} e^{-iu\lambda}d\nu(\lambda)=\frac{1}{\sqrt{2\pi}}\int_{\mathbb{R}} e^{-iu\lambda}\alpha(\lambda)\cdot\alpha(\lambda)dm(\lambda)\\&=\frac{1}{\sqrt{2\pi}}\int_{\mathbb{R}}\alpha(\lambda-u)\overline{\alpha(\lambda)}dm(\lambda)=\frac{1}{\sqrt{2\pi}}\int_{\mathbb{R}}\alpha(\lambda-u)\alpha(\lambda)dm(\lambda).\end{aligned}\tag{20}$$

The third equality is assured by Lemma 1.

Divide $\mathbb{R}$ into $\mathbb{R}_+$ and $\mathbb{R}_0$ defined as $\mathbb{R}_+=\{t\in\mathbb{R}|\alpha(t)>0\}$, $\mathbb{R}_0=\{t\in\mathbb{R}|\alpha(t)=0\}$.

As in Theorem 1, we introduce the orthogonal measure $\eta(S,\omega):\mathscr{B}(\mathbb{R})\times\Omega\to\mathbb{C}$, two functions $\alpha_1(\cdot)$ and $\alpha_2(\cdot)$, and another orthogonal measure $\gamma(S,\omega)$. We can construct η and γ so as to be

$$\nu_\eta(S)=m(S),\quad \nu_\gamma=m(S\cap\mathbb{R}_+)\quad\text{for}\quad S\in\mathscr{B}^*(\mathbb{R})\quad\text{(see section 4)}.$$

If we define a weakly stationary process $X'(t,\omega)$ by

$$X'(t,\omega)=\frac{1}{\sqrt[4]{2\pi}}\int_{\mathbb{R}}\alpha(\lambda-t)\gamma(d\lambda,\omega),$$

then the covariance $\rho'(u)$ of $X'(t,\omega)$ is[18]

$$\begin{aligned}\rho'(u)&=\mathbb{E}X'(t+u,\omega)\overline{X'(t,\omega)}\\&=\frac{1}{\sqrt{2\pi}}\mathbb{E}\int_{\mathbb{R}}\alpha(\lambda-(t+u))\gamma(d\lambda,\omega)\int_{\mathbb{R}}\overline{\alpha(\lambda-t)\gamma(d\lambda,\omega)}\end{aligned}\tag{21}$$

[18]The third line of (21)

$$=\frac{1}{\sqrt{2\pi}}\left\{\int_{\mathbb{R}_+}\alpha(\lambda-(t+u))\alpha(\lambda-t)dm(\lambda)+\int_{\mathbb{R}_0}\underbrace{\alpha(\lambda-(t+u))\alpha(\lambda-t)}_{(\dagger)}d\nu_\gamma\right\}$$

$$=\frac{1}{\sqrt{2\pi}}\left\{\int_{\mathbb{R}}(\dagger)dm(\lambda)-\int_{\mathbb{R}_0}(\dagger)dm(\lambda)+\int_{\mathbb{R}_0}(\dagger)d\nu_\gamma\right\}$$

$$= \frac{1}{\sqrt{2\pi}} \int_{\mathbb{R}} \alpha(\lambda - (t+u))\alpha(\lambda - t) d\nu_\gamma(\lambda) \qquad \text{(by D-K formula)}$$
$$= \frac{1}{\sqrt{2\pi}} \int_{\mathbb{R}} \alpha(\lambda - u)\alpha(\lambda) dm(\lambda).$$

By (20) and (21), we get $\rho = \rho'$. That is, the spectral density function of $X'(t, \omega)$ is $p(\cdot)$.

(ii)⇒(i): Conversely, assume (ii). The covariance $\rho(u)$ of

$$X(t, \omega) = \int_{\mathbb{R}} w(\lambda - t)\gamma(d\lambda, \omega)$$

is given by

$$\rho(u) = \int_{-\infty}^{\infty} w(\lambda - u)\overline{w(\lambda)} dm(\lambda)$$

as in the calculation of $\rho'(\cdot)$ above.

Then we get, again by Lemma 1,

$$\rho(u) = \int_{\mathbb{R}} w(\lambda - u)\overline{w(\lambda)} dm(\lambda)$$
$$= \int_{\mathbb{R}} |w(\lambda)|^2 e^{-iu\lambda} dm(\lambda)$$
$$= \frac{1}{\sqrt{2\pi}} \int_{\mathbb{R}} \sqrt{2\pi} |w(\lambda)|^2 e^{-iu\lambda} dm(\lambda).$$

Finally, if we define $p(\lambda) = \sqrt{2\pi}|w(\lambda)|^2$, $p(\lambda)$ is the spectral density function of $X(t, \omega)$. **Q.E.D.**

$$= \frac{1}{\sqrt{2\pi}} \Big\{ \int_{\mathbb{R}} (\dagger) dm(\lambda)$$
$$- \int_{\substack{\lambda \in \mathbb{R}_0 \\ \lambda - t \in \mathbb{R}_+}} (\dagger) dm(\lambda) - \int_{\substack{\lambda \in \mathbb{R}_0 \\ \lambda - t \in \mathbb{R}_0}} (\dagger) dm(\lambda) + \int_{\substack{\lambda \in \mathbb{R}_0 \\ \lambda - t \in \mathbb{R}_+}} (\dagger) d\nu_\gamma(\lambda) + \int_{\substack{\lambda \in \mathbb{R}_0 \\ \lambda - t \in \mathbb{R}_0}} (\dagger) d\nu_\gamma(\lambda) \Big\}$$
$$= \frac{1}{\sqrt{2\pi}} \Big\{ \int_{\mathbb{R}} (\dagger) dm(\lambda)$$
$$- \int_{(\mathbb{R}_0 - t) \cap \mathbb{R}_+} \alpha(\lambda' - u)\alpha(\lambda') dm(\lambda') + \int_{(\mathbb{R}_0 - t) \cap \mathbb{R}_+} \alpha(\lambda' - u)\alpha(\lambda') d\nu_\gamma(\lambda') \Big\}.$$

The last two terms cancel out. So we obtain (21).

8 Conclusion

A (measurable) weakly stationary process $X(t, \omega)$ is τ-periodic if and only if its spectral measure ν concentrates on a countable set in $\mathbb{T}$ or $\mathbb{R}$ such that the distance of any adjacent two points is some multiple of $2\pi/\tau$ (cf. Proposition 6, 6').

$X(t, \omega)$ is almost periodic if and only if ν is discrete (cf. Proposition 7, 7').

Therefore if $X(t, \omega)$ is periodic or almost periodic, the spectral measure ν can never be absolutely continuous with respect to the Lebesgue measure.

Now under what conditions does the spectral measure have the density function? The answers were given in Theorems 1 and 2. ν is absolutely continuous with respect to the Lebesgue measure if and only if $X(t, \omega)$ is a moving average process of white noise.

So we can conclude that *any moving average process can neither be periodic nor almost periodic.*

However we know that the set $\mathfrak{M}_+^d(\mathbb{R}) = \left\{ \sum_{j=1}^{n} \alpha_j \delta_{x_j} \middle| \alpha_j \in \mathbb{R}, x_j \in \mathbb{R}, n \in \mathbb{N} \right\}$ of all the discrete positive Radon measures is w^*-dense in $\mathfrak{M}_+(\mathbb{R})$, the set of all the positive Radon measures (cf. Billingsley [1], Malliavin [22] Chap. II, §6).

The following Theorem immediately follows from this observation.

Theorem 3. *Let $X(t, \omega)$ be a measurable and weakly stationary process on $\mathbb{R} \times \Omega$ with the spectral measure ν. Then there exists a sequence of almost periodic weakly stationary process $X^k(t, \omega)$ with the spectral measure ν^k such that ν^k converges to ν in the w^*-topology.*

Proof. Since $\mathfrak{M}_+^d(\mathbb{R})$ is w^*-dense in $\mathfrak{M}_+(\mathbb{R})$ (metrizable), there exists a sequence $\nu^k \in \mathfrak{M}_+^d(\mathbb{R})$ which w^*-converges to ν. By Proposition 5, there is a measurable and weakly stationary process $X^k(t, \omega)$, the spectral measure of which is exactly equal to ν^k. $X^k(t, \omega)$ is almost periodic because $\nu^k \in \mathfrak{M}_+^d(\mathbb{R})$. **Q.E.D.**

A similar result also holds true for discrete time case. (In this case $\mathbb{R}$ should be replaced by $\mathbb{T}$ as usual.)

Theorem 3 tells us that *any weakly stationary process $X(t, \omega)$ can be approximated by a sequence $\{X^k(t, \omega)\}$ of almost periodic weakly stationary processes* in the sense that the sequence of the spectral measures of $X^k(t, \omega)$ converges to the spectral measure of $X(t, \omega)$ in the w^*-topology.

Appendix

We reviewed, in the first few sections, some basic concepts and results to be required in our main concerns (Theorems 1, 2, and 3). I expect this exposition to be a small help for the readers who are not familiar with them. There can be no communications without common language. Of course, these materials are, more or less, known to every mathematician working in this discipline. That is why we just stated necessary facts and dropped proofs in the text.

However it does not seem to be a waste of time to give here complete proofs of selected three propositions (Propositions 5', 6' and 7') as exceptions, taking account of their indispensable roles in our problem.

Proof of Proposition 5'. The proof is basically due to Kawata [19] pp. 56–57.[19] If we define $\theta : \mathscr{B}(\mathbb{R}) \to \mathbb{R}$ by

$$\theta(E) = \frac{\nu(E)}{\nu(\mathbb{R})}, \quad E \in \mathscr{B}(\mathbb{R}),$$

then θ is a Radon probability measure. There exist independent real random variables $Z(\omega)$ and $Y(\omega)$ defined on some probability space $(\Omega, \mathscr{E}, P)$ such that the distribution of $Z(\omega)$ is equal to θ, $\mathbb{E}Y(\omega) = 0$, and finally $\mathbb{E}Y(\omega)^2 = (1/\sqrt{2\pi})\nu(\mathbb{R})$. [20] Define a stochastic process $X(t, \omega)$ by

$$X(t, \omega) = Y(\omega)e^{iZ(\omega)t}.$$

Then $\mathbb{E}X(t, \omega)$ and $\rho(t + u, t)$ are calculated as follows.[21]

$$\begin{aligned} \mathbb{E}X(t, \omega) &= \mathbb{E}Y(\omega)\mathbb{E}e^{iZ(\omega)t} = 0 \quad \text{(by independence)}, \\ \rho(t + u, t) &= \mathbb{E}X(t + u, \omega)\overline{X(t, \omega)} = \mathbb{E}[y(\omega)^2 e^{iZ(\omega)u}] \\ &= \frac{1}{\sqrt{2\pi}}\nu(\mathbb{R}) \int_{\mathbb{R}} e^{i\lambda u} d\theta(\lambda) = \frac{1}{\sqrt{2\pi}} \int_{\mathbb{R}} e^{i\lambda u} d\nu(\lambda). \end{aligned}$$

Thus the covariance $\rho(t+u, t)$ of $X(t, \omega)$ depends only upon u, and it is expressed as the Fourier transform of ν. Since the Fourier transform of ν is uniformly continuous, $X(t, \omega)$ is strongly continuous on $\mathbb{R}$ by Proposition 1. The measurability of $X(t, \omega)$ is obvious. **Q.E.D.**

[19]However we have to be more careful about a couple of subtle reasonings. (a) $(\Omega, \mathscr{E}, P)$ can not be fixed a priori. It must be chosen suitably. (b) A justification must be given for the construction of independent random variables.

[20]Let $\Phi_1, \Phi_2, \cdots$ be a sequence of Borel probability measures on $\mathbb{R}$. Then there exists a sequence of independent random variables defined on some probability space $(\Omega, \mathscr{E}, P)$, the distributions of which are $\Phi_1, \Phi_2, \cdots$ (cf. Itô [15] p. 68).

[21]Let $X_1, X_2, \cdots$ be a sequence of independent real-valued random variables, and $g_1, g_2, \cdots : \mathbb{R} \to \mathbb{C}$ Borel measurable functions. Then $Y_1 = g_1(X_1), Y_2 = g_2(X_2), \cdots$ are also independent random variables (cf. Itô [15] p. 66). So $Y(\omega)$ and $e^{iZ(\omega)t}$ in the text are independent.

Proof of Proposition 6'[22]. (i)⇒ (ii): Assume that $\rho(u)$ is τ-periodic. We then prove that

$$\mathbb{E}|X(t+\tau,\omega)-X(t,\omega)|^2=0$$

which is equivalent to (ii). By direct computation, we have

$$\begin{aligned}
&\mathbb{E}|X(t+\tau,\omega)-X(t,\omega)|^2\\
&=\mathbb{E}|X(t+\tau,\omega)|^2+\mathbb{E}|X(t,\omega)|-2\mathscr{R}e\,\mathbb{E}X(t+\tau,\omega)\overline{X(t,\omega)}\\
&=2\rho(0)-2\mathscr{R}e\rho(\tau)\\
&=2(\rho(0)-\mathscr{R}e\rho(\tau))\\
&=0 \quad \text{(by (i))}.
\end{aligned}$$

(ii)⇒ (i): Assume (ii). then we have

$$\begin{aligned}
|\rho(u+\tau)-\rho(u)|^2&=|\mathbb{E}[X(u+\tau,\omega)\overline{X(0,\omega)}-X(u,\omega)\overline{X(0,\omega)}]|^2\\
&=0. \quad \text{(by (ii))}
\end{aligned}$$

(i)⇒ (iii): If $\rho(u)$ is τ-periodic, then we have

$$0=2\rho(0)-\rho(\tau)-\rho(-\tau)=\frac{2}{\sqrt{2\pi}}\int_{\mathbb{R}}(1-\cos t\tau)d\nu(t).$$

Taking account of $1-\cos t\tau \geqq 0$, we must have $\nu(E)=0$ for any $E\in\mathscr{B}(\mathbb{R})$ such that

$$E\cap\{t\in\mathbb{R}|1-\cos t\tau=0\}=\emptyset,$$

which is equivalent to

$$E\cap\{2k\pi/\tau|k\in\mathbb{Z}\}=\emptyset.$$

(iii)⇒ (i): By definition of the spectral measure,

$$\rho(u)=\frac{1}{\sqrt{2\pi}}\int_{\mathbb{R}}e^{-iut}d\nu(t).$$

[22]We follow Kawata [19] pp. 75–76.

So putting $a_k = \nu(\{2\pi k/\tau\})$ ($k \in \mathbb{Z}$), we obtain, by (iii), that

$$\rho(u) = \frac{1}{\sqrt{2\pi}} \sum a_k e^{-iu\cdot 2\pi k/\tau}.$$

This is clearly τ-periodic. **Q.E.D.**

The following well-known lemma[23] will be used in the course of the proof of Proposition 7'.

Lemma. *Let μ be a Radon measure on $\mathbb{R}$. If the Fourier transform $\hat{\mu}$ is almost periodic, then μ is discrete.*

Proof of Proposition 7'[24]. (i)⇒(ii): Assume that $\rho(\mu)$ is almost periodic. Then we have

$$\begin{aligned}\sup_{t\in\mathbb{R}} \mathbb{E}|X(t+\tau,\omega) - X(t,\omega)|^2 &= 2[\rho(0) - \mathscr{R}e\rho(\tau)] \\ &= 2\mathscr{R}e[\rho(0) - \rho(\tau)] \leqq 2|\rho(0) - \rho(\tau)| \qquad (1) \\ &\leqq 2\sup_{u\in\mathbb{R}} |\rho(u+\tau) - \rho(u)|.\end{aligned}$$

Since $\rho(\cdot)$ is almost periodic, there exists a number $\Lambda(\varepsilon/2,\rho) > 0$, for each $\varepsilon > 0$, such that any interval the length of which is $\Lambda(\varepsilon/2,\rho)$ contains an $\varepsilon/2$- almost period, τ. Since

$$\begin{aligned}&\{\tau \in \mathbb{R} | \sup_{u\in\mathbb{R}} |\rho(u+\tau) - \rho(u)| < \frac{\varepsilon}{2}\} \\ &\subset \{\tau \in \mathbb{R} | \sup_{t\in\mathbb{R}} \mathbb{E}|X(t+\tau,\omega) - X(t,\omega)|^2 < \varepsilon\}\end{aligned}$$

by (1), we see that (ii) is satisfied by setting $\Gamma(\varepsilon, X) = \Lambda(\varepsilon/2,\rho)$.

(ii)⇒(i): Assume (ii). By a simple calculation, we obtain the inequality

$$\begin{aligned}|\rho(u+\tau) - \rho(u)|^2 &= |\mathbb{E}[X(u+\tau,\omega)\overline{X(0,\omega)} - X(u,\omega)\overline{X(0,\omega)}]|^2 \\ &\leqq \mathbb{E}|X(u+\tau,\omega) - X(u,\omega)|^2 \mathbb{E}|X(0,\omega)|^2 \\ &\qquad \text{(by Schwarz inequality).}\end{aligned}$$

It follows that

$$|\rho(u+\tau) - \rho(u)| \leqq [\mathbb{E}|X(t+\tau,\omega) - X(u,\omega)|^2]^{1/2} \rho(0)^{1/2}.$$

---

[23]Katznelson [16] p. 197.

[24]The proof of (i)⇔(ii) is due to Kawata [19] pp. 80–82.

Hence we have

$$\left\{\tau \in \mathbb{R} \,\middle|\, \sup_{u\in\mathbb{R}} \mathbb{E}|X(u+\tau,\omega)-X(u,\omega)|^2 < \frac{\varepsilon^2}{\rho(0)}\right\}$$
$$\subset \{\tau \in \mathbb{R} | \sup_{u\in\mathbb{R}} |\rho(u+\tau)-\rho(u)| < \varepsilon\}.$$

Thus (i) holds true by setting $\Lambda(\varepsilon, \rho) = \Gamma(\varepsilon^2/\rho(0), X)$.

(i)⇒ (iii): This is a direct consequence of the lemma. The covariance function ρ is represented by the Fourier transform of some positive Radon measure ν on $\mathbb{R}$ (Proposition 4'). Since $\rho(u) = \hat{\nu}(u)$ is almost periodic, ν must be discrete by the lemma.

(iii)⇒ (i): Assume that the spectral measure ν of X is discrete, say

$$\nu = \sum_{n=1}^{\infty} a_n \delta_{\xi_n} \quad (\delta_{\xi_n} : \text{Dirac measure}).$$

Then the covariance can be expressed as

$$\rho(u) = \frac{1}{\sqrt{2\pi}} \int_{\mathbb{R}} e^{-i\lambda u} d\nu(\lambda) = \frac{1}{\sqrt{2\pi}} \sum_{n=1}^{\infty} a_n e^{-i\xi_n u}. \tag{2}$$

Since $a_n \geqq 0$, the series (2) is absolutely and so uniformly convergent. Thus $\rho(u)$ is the uniform limit of trigonometric polynomials,[25] and hence almost periodic.

Q.E.D.

References

1. Billingsley P (1968) Convergence of probability measures. Wiley, New York
2. Bochner S (1932) Vorlesungen über Fouriersche Integrale. Akademische Verlagsgesellschaft, Leipzig
3. Bochner S (1933) Monotone Funktionen, Stieltjessche Integrale und harmonische Analyse. Math Ann 108:378–410
4. Bohr H (1932) Fastperiodische Funktionen. Springer, Berlin
5. Brémaud P (2014) Fourier analysis and stochastic processes. Springer, Cham

[25] A function of the form

$$f(x) = \sum_{j=1}^{n} a_j e^{-i\xi_j u} \quad (\xi_j \in \mathbb{R})$$

is called a trigonometric polynomial. Any trigonometric polynomial is almost periodic.

6. Carleson L (1966) On convergence and growth of partial sums of Fourier series. Acta Math 116:135–157
7. Cramér H (1940) On the theory of stationary random processes. Ann Math 41:215–230
8. Crum MM (1956) On positive-definite functions. Proc Lond Math Soc (3) 6:548–560
9. Doob JL (1953) Stochastic processes. Wiley, New York
10. Doob JL (1949) Time series and harmonic analysis. In: Proceedings of Berkeley Symposium on Mathematical Statistics and Probability. University of California Press, Berkeley, pp 303–393
11. Frisch R (1933) Propagation problems and inpulse problems in dynamic economics. In: Economic essays in Honor of Gustav Cassel. Allen and Unwin, London
12. Granger CWJ, Newbold P (1986) Forecasting economic time series, 2nd edn. Academic, Orlando
13. Hamilton JD (1994) Time series analysis. Princeton University Press, Princeton
14. Herglotz G (1911) Über Potenzreihen mit positiven reellen Teil in Einheitskreis. Berichte Verh Säcks. Akad Wiss Leibzig Math Phys Kl 63:501–511
15. Itô K (1953) Probability theory. Iwanami, Tokyo (in Japanese)
16. Katznelson Y (2004) An introduction to harmonic analysis, 3rd edn. Cambridge University Press, Cambridge
17. Kawata T (1966) On the Fourier series of a stationary stochastic process I. Z Wahrsch Verw Gebiete 6:224–245
18. Kawata T (1969) On the Fourier series of a stationary stochastic process II. Z Wahrsch Verw Gebiete 13:25–38
19. Kawata T (1985) Stationary stochastic processes. Kyoritsu, Tokyo (in Japanese)
20. Lax PD (2002) Functional analysis. Wiley, New York
21. Loomis LH (1960) The spectral characterization of a class of almost periodic functions. Ann Math 72:362–368
22. Malliavin P (1995) Integration and probability. Springer, New York
23. Maruyama G (1949) The harmonic analysis of stationary stochastic processes. Mem Fac Sci Kyushu Univ Ser A 4:45–106
24. Naimark MA (1972) Normed algebras. Wolters Noordhoff, Groningen
25. Rudin W (1962) Fourier analysis on groups. Wiley, New York
26. Sargent TJ (1979) Macroeconomic theory. Academic, New York
27. Schwartz L (1966) Théorie des distributions. Hermann, Paris
28. Slutsky E (1937) Alcune proposizioni sulla teoria delle funzioni aleatorie. Giorn Inst Ital degli Attuari 8:193–199
29. Slutsky E (1938) Sur les fonctions aléatoires presque périodiques et sur la décomposition des fonctions aléatoires stationnaires en composantes. Actualités Sci Ind 738:38–55
30. Slutsky E (1937) The summation of random causes as the source of cyclic processes. Econometrica 5:105–146
31. Wold H (1953) A study in the analysis of stationary time series, 2nd edn. Almquist and Wicksell, Uppsala

The 6th Conference on Mathematical Analysis in Economic Theory

Date: January 26(Mon)– 29(Thu), 2015
Venue: Lecture Hall, East Research Building,
Keio University
2-15-45 Mita, Minato-ku, Tokyo 108-8345, JAPAN

organized by The Japanese Society for Mathematical Economics
cosponsored by Keio Economic Society, The Oak Society, Inc.

Programme

January 26 (Monday) *Speaker

Morning

	Chair: Takuji Arai (Keio University)
9:00–10:00	**Keita Owari** (The University of Tokyo) On the Lebesgue property and related regularities of monotone convex functions on Orlicz spaces
10:00–11:00	**Shigeo Kusuoka** (The University of Tokyo) Expectation of diffusion process with absorbing boundary
11:10–12:10	**Robert Anderson*** (UC Berkeley), L.R. Goldberg, N. Gunther The cost of financing leveraged US equity through futures

S. Kusuoka, T. Maruyama (eds.), *Advances in Mathematical Economics Volume 20*, Advances in Mathematical Economics,
DOI 10.1007/978-981-10-0476-6

Afternoon

	Chair: Ryozo Miura (Hitotsubashi University)
13:30–14:30	**Takuji Arai** (Keio University) Local risk-minimization for Barndorff-Nielsen and Shephard models
14:30–15:30	**Katsumasa Nishide** (Yokohama National University) Heston-type stochastic volatility with a Markov switching regime
	Chair: Ken Urai (Osaka University)
16:00–17:00	**Hisashi Inaba** (The University of Tokyo) Recent developments of the basic reproduction number theory in population dynamics
17:00–18:00	**Takashi Suzuki*** (Meiji Gakuin University), Nobusumi Sagara Exchange economies with infinitely many commodities and a saturated measure space of consumers

January 27 (Tuesday)

Morning

	Chair: Takashi Suzuki (Meiji Gakuin University)
9:00–10:00	**Yuhki Hosoya** (Kanto Gakuin University) The NLL axiom and integrability theory
10:00–11:00	**Nobusumi Sagara** (Hosei University) An indirect method of nonconvex variational problems in Asplund spaces : the case for saturated measure spaces
11:10–12:10	**Ali Khan*** (Johns Hopkins University), Yongchao Zhang On pure-strategy equilibria in games with correlated information

Afternoon

	Chair: Hidetoshi Komiya (Keio University)
13:30–14:30	**Hiroyuki Ozaki** (Keio University) Upper-convergent dynamic programming
14:30–15:30	**Vladimir Tikhomirov** (Moscow State University) "Problems of the theory of extremal problems and applications"
	Chair: Ali Khan (Johns Hopkins University)
16:00–17:00	**Alexander Ioffe** (Israel Institute of Technology) On curves of descent
17:00–18:00	**Arturo Kohatsu-Higa** (Ritsumeikan University) The probabilistic parametrix method as a simulation method (tentative)

January 28 (Wednesday)

Morning

	Chair: Kazuya Kamiya (The University of Tokyo)
9:00–10:00	**Yoshiyuki Sekiguchi** (Tokyo University of Marine Science and Technology) Real algebraic methods in optimization
10:00–11:00	**Ronaldo Carpio** (University of International Business & Economics) Fast Bellman iteration: An application of Legendre-Fenchel duality to infinite-horizon dynamic programming in discrete time
11:10–12:10	**Takashi Kamihigashi** (Kobe University) Extensions of Fatou's lemma and the dominated convergence theorem

Poster Session

12:10–13:30	**Takeshi Ogawa** (Hiroshima Shudo University) Intermediate goods with Leontief 's model and joint production with activity analysis in Ricardian comparative advantage **Satoru Kageyama** (Osaka University), Ken Urai Fiscal stabilization policy in a Phillips model with unstructured uncertainty **Hiromi Murakami** (Osaka University), Ken Urai Replica core equivalence theorem: an extension of Debreu-Scarf limit theorem to double infinity monetary economies **Ryonfun Im** (Kobe University), Takashi Kamihigashi An equilibrium model with two types of asset bubbles

Afternoon

	Chair: Shinichi Suda (Keio University)
13:30–14:30	**Makoto Hanazono** (Nagoya University) Procurement auctions with general price-quality evaluation
14:30–15:30	**Chiaki Hara*** (Kyoto University), Kenjiro Hirata Dynamic inconsistency in pension fund management
	Chair: Shigeo Kusuoka (The University of Tokyo)
16:00–17:00	**Chia-Hui Chen** (Kyoto University) A tenure-clock problem: evaluation, deadline, and up-or-out
17:00–18:00	**Nozomu Muto*** (Yokohama National University), Shin Sato Bounded response and Arrow's impossibility
18:30–21:00	Reception at Tsunamachi Mitsui Club

January 29 (Thursday) Satellite Session

Morning

	Chair: Hiroyuki Ozaki (Keio University)
9:00–9:30	**Masayuki Yao** (Keio University) Recursive utility and dynamic programming under the assumption of upper convergence: an order-theoretic approach
9:30–10:00	**Chaowen Yu** (Keio University) Locally robust mechanism design
10:00–10:30	**Hiromi Murakami*** (Osaka University), Ken Urai Replica core equivalence theorem: an extension of Debreu-Scarf limit theorem to double infinity monetary economies
11:00–11:30	**Kohei Shiozawa*** (Osaka University), Ken Urai A generalization of social coalitional equilibrium for multiple coalition structures
11:30–12:00	**Takeshi Ogawa** (Hiroshima Shudo University) Intermediate goods with Leontief 's model and joint production with activity analysis in Ricardian comparative advantage

Robert Anderson

Takashi Suzuki

Yuhki Hosoya

Nobusumi Sagara

Ali Khan

Chiaki Hara

Vladimir Tikhomirov, Toru Maruyama, Alexander Ioffe

Saying “Good-bye" at the reception (Tsunamachi Mitsui Club)

Index

S. Kusuoka, T. Maruyama (eds.), *Advances in Mathematical Economics Volume 20*, Advances in Mathematical Economics,
DOI 10.1007/978-981-10-0476-6

Printed in the United States
By Bookmasters